NIMS Machining Level 1
Study Guide

Andrew J. Klein

CENGAGE
Learning·

Australia • Brazil • Mexico • Singapore • United Kingdom • United States

NIMS Machining Level 1 Study Guide
Andrew J. Klein

SVP, GM Skills & Global Product Management: Dawn Gerrain

Product Director: Matthew Seeley

Associate Product Manager: Nicole Robinson

Senior Director, Development: Marah Bellegarde

Senior Product Development Manager: Larry Main

Associate Content Developer: Jenn Alverson

Product Assistant: Maria Garguilo

Vice President, Marketing Services: Jennifer Ann Baker

Marketing Director: Michele McTighe

Senior Production Director: Wendy Troeger

Production Director: Andrew Crouth

Content Project Management and Art Direction: Lumina Datamatics, Inc.

Cover image(s): Andy Livin

Unless otherwise noted, all items © 2017 Cengage Learning.

© 2017 Cengage Learning

WCN: 01-100-101

For product information and technology assistance, contact us at **Cengage Learning Customer & Sales Support, 1-800-354-9706.**

For permission to use material from this text or product, submit all requests online at **www.cengage.com/permissions.** Further permissions questions can be e-mailed to **permissionrequest@cengage.com.**

Library of Congress Control Number: 2015940562

ISBN-13: 978-1-285-42277-0

Cengage Learning
20 Channel Center Street
Boston, MA 02210
USA

Cengage Learning is a leading provider of customized learning solutions with employees residing in nearly 40 different countries and sales in more than 125 countries around the world. Find your local representative at **www.cengage.com**.

Cengage Learning products are represented in Canada by Nelson Education, Ltd.

To learn more about Cengage Learning, visit **www.cengage.com**

Purchase any of our products at your local college store or at our preferred online store **www.cengagebrain.com**.

Notice to the Reader

Publisher does not warrant or guarantee any of the products described herein or perform any independent analysis in connection with any of the product information contained herein. Publisher does not assume, and expressly disclaims, any obligation to obtain and include information other than that provided to it by the manufacturer. The reader is expressly warned to consider and adopt all safety precautions that might be indicated by the activities described herein and to avoid all potential hazards. By following the instructions contained herein, the reader willingly assumes all risks in connection with such instructions. The publisher makes no representations or warranties of any kind, including but not limited to, the warranties of fitness for particular purpose or merchantability, nor are any such representations implied with respect to the material set forth herein, and the publisher takes no responsibility with respect to such material. The publisher shall not be liable for any special, consequential, or exemplary damages resulting, in whole or part, from the readers' use of, or reliance upon, this material.

Table of Contents

Preface

This guide has been developed to aid machining students and working professionals in achieving high levels of success on the National Institute for Metalworking Skills (NIMS) certification exams. After receiving technical training, practicing, and demonstrating the competencies assessed by a specific NIMS Level I certification exam, this study guide will help you determine your level of readiness for the actual NIMS certification exam. The alignment of the test questions in this study guide to the NIMS certification exams will assist the testing candidate in determining his or her strengths and weaknesses in the skill area being assessed. Each question and the correct answer are identified and clarified in the Explanation section at the end of the corresponding sample test. These explanations will provide the testing candidate insight into the knowledge and skill area being assessed and serve as an extension of the classroom, lab, and on-the-job training previously received.

The study guide begins with an introduction to NIMS and the Level I certifications. This chapter details the registration process, the performance exams, and the requirements for the online theory exams. In addition, this chapter explains the format of the practice tests and concludes with test-taking strategies to increase success when taking a NIMS certification exam. The subsequent chapters contain the practice tests, answer keys, and question explanations for the 11 NIMS Level I certification exams. A glossary of terms is also included as the final section.

ABOUT THE AUTHOR

Andrew J. Klein, Western Montgomery Career and Technology Center
Limerick, Pennsylvania

Andrew Klein continues to build on 22 years of diverse machining and metalworking experience. He holds dual AAS degrees in machine tool technology and automated manufacturing from Pennsylvania College of Technology, a BS in business administration from Alvernia University, a MEd in career and technical education from Temple University, and a vocational education II teaching certification in Pennsylvania. He has earned all of the Level I NIMS machining certifications, the NIMS Level II EDM certification, and journey person tool and die certification through the State of Pennsylvania. Andrew was the recipient of the 2012 ACTE/NIOSH national lab safety award.

COVER IMAGES ACKNOWLEDGEMENT

The cover images are courtesy of Andy Livin, Kirkwood Community College, Cedar Rapids, Iowa.

Introduction to NIMS*

1

"To strengthen American manufacturing by building a globally competitive metalworking workforce."

1.1 ABOUT NIMS

The National Institute for Metalworking Skills (NIMS) was formed in 1995 by the metalworking trade associations to develop and maintain a globally competitive American workforce. NIMS sets skills standards for the industry, certifies individual skills against the standards, and accredits training programs that meet NIMS quality requirements. NIMS operates under rigorous and highly disciplined processes as the only developer of American National Standards for the nation's metalworking industry accredited by the American National Standards Institute (ANSI).

NIMS has a stakeholder base of over 6000 metalworking companies. The major trade associations in the industry—the Association for Manufacturing Technology, the National Tooling & Machining Association, the Precision Machine Products Association, the Precision Metalforming Association, and the Tooling and Manufacturing Association—have invested over $7.5 million in private funds for the development of the NIMS standards and its credentials. The associations also contribute annually to sustain NIMS operations and are committed to the upgrading and maintenance of the standards.

NIMS has developed skills standards in 24 operational areas covering the breadth of metalworking operations including metalforming (stamping, press brake, roll forming, laser cutting) and machining (machining, tool and diemaking, mold making, screw machining, machine building and machine maintenance, service and repair). The standards range from entry (Level I) to a master level (Level III). All NIMS standards are industry written and industry validated and are subject to regular periodic reviews under the procedures accredited and audited by ANSI.

NIMS certifies individual skills against the national standards. The NIMS credentialing program requires that the candidate meet both performance and theory requirements. Both the performance and knowledge examinations are industry designed and industry piloted. There are 52 distinct NIMS skill certifications. The industry uses the credentials to recruit, hire, place, and promote individual workers. The credentials are often incorporated as completion requirements and as a basis for articulation among training programs.

---

*This text is adapted from https://www.nims-skills.org/.

1.2 NIMS LEVEL I MACHINING CERTIFICATIONS

The NIMS Level I machining certifications are designed to meet entry-level requirements for on-the-job skills:

Measurement, Materials, and Safety

Job Planning, Benchwork, and Layout

Manual Milling Skills I

Turning Operations: Turning between Centers

Turning Operations: Turning Chucking Skills

Grinding Skills I

Drill Press Skills I

CNC Turning: Programming Setup and Operations

CNC Milling: Programming Setup and Operations

CNC Turning: Operations

CNC Milling: Operations

1.3 NIMS CERTIFICATION PROCESS

Step One—Candidate Registration

Candidate registration is completed on the NIMS Web site and requires a $40 registration fee per person. Registration is good for life, so the $40 fee is assessed only once per person. Only individuals seeking NIMS credentials are required to register.

Students/trainees may complete the NIMS performance exam prior to registering; however, registration must be completed prior to taking an online theory exam.

Registration may be completed by the student/trainee or on his or her behalf by a sponsor or other employee.

Registering a Student/Trainee as a Candidate

1. Go to www.nims-skills.org and click **Candidate Registration** on the menu to the left.
2. Select your organization and the student/trainee's sponsor from the first two drop-down lists.
3. Complete the rest of the form to reflect the student/trainee's home address and contact information. If this information is unavailable at the time of registration, simply enter the address and contact information of the school or company where training occurs.
4. Enter a valid e-mail address in the last box of the form. NIMS staff will send two messages to this e-mail address to confirm this registration, so enter a valid address. Sponsors may enter

their own e-mail address in this box. The following two e-mails will arrive within 24 hours of registration:

a. A receipt for payment of the $40 registration fee.
b. A candidate username and temporary password. The student/trainee will use these to log into the NIMS online testing center at www.nims-skills.org, where he or she will take Online Theory Exams.

Step Two—The Candidate Completes a Performance Exam

Performance Exams are required for all NIMS credentials—except Measurement, Materials, and Safety (MMS). If you are seeking this MMS credential, you may skip to step three.

Most Machining Level I and II Performance Exams require the candidate to machine a part based on a NIMS-provided print. Part prints are found on the Resources page online or on the NIMS Tools & Documents CD.

Some Machining Level I, II, and II Performance Exams require the candidate to complete a Credentialing Achievement Record (CAR). This is a comprehensive checklist of skills and is to be completed under sponsor supervision. CARs are found on the Resources page online or on the NIMS Tools & Documents CD.

All other NIMS credentials require a CAR as a Performance Exam. This includes Metalforming Level I; Stamping Levels II and III; Press Brake Levels II and III; Slide Forming Levels II and III; Screw Machining Levels II and III; Machine Building Levels II and III; Machine Maintenance, Service, and Repair Levels II and III; and Diemaking Levels II and III.

Level I Performance Exams do not have a time limit; however, most Level II and III Performance Exams do have a time limit, which can be found in the Performance Guide to Machining Level II or in the CAR.

The Performance Exam Process

1. Sponsors should distribute part prints or CAR to candidates prior to starting the Performance Exam.
2. Candidates are to complete the Performance Exam under the supervision of a NIMS-registered sponsor.

Once a candidate completes the part—

If a candidate has completed the two parts required for the Job Planning, Benchwork, and Layout credential, then the sponsor may inspect both. If the parts are within 100% tolerance of the prints, the sponsor will then complete the top and middle portion of the performance affidavit and then fax that affidavit to NIMS staff at (703) 352-4991 or e-mail it to support@nims-skills.org.

If a candidate has completed a part for any other machining credential, the sponsor is to complete the top and middle portion of the Performance Affidavit for that specific candidate and then make arrangements with his or her MET-TEC Committee to have that part inspected. For more information on MET-TEC Committees, consult the Complete Guide to NIMS Credentialing Program available on the NIMS Web site. If a part passes inspection by no less than two members of a MET-TEC Committee, then both inspectors should complete the bottom portion of the affidavit and fax that affidavit to NIMS staff at (703) 352-4991 or e-mail it to support@nims-skills.org.

Submitting affidavits alerts NIMS staff that the candidate has completed and passed the Performance Exam. At this point a staff member will make the corresponding Online Theory Exam available.

Once a candidate completes a CAR—

The sponsor is to complete the affidavit of successful completion, which is found at the end of the CAR booklet, and then fax or e-mail that affidavit to NIMS staff at (703) 352-4991 or by e-mail at support@nims-skills.org.

Submitting the affidavit will alert the NIMS staff that the candidate has completed and passed the Performance Exam. At this point a staff member will make the corresponding Online Theory Exam available.

Step Three—The Candidate Takes an Online Theory Exam

The last step to earning any NIMS credential is to take the Online Theory Exam while under proctor supervision. For information on selecting and registering a proctor, please see www.nims-skills.org.

Sponsors are to work with their proctor to select a time and place for candidates' online test. Online Theory Exams are priced at $35 for Level I credentials and at $50 for Level II and III credentials. Training programs that are accredited by NIMS receive a 20% discount, making fees $28 and $40, respectively. Each Online Theory Exam must be completed within 90 minutes. A time counter is located in the top-right corner of the screen once the exam is started.

A pass or fail grade is displayed on-screen after submitting answers for scoring. If the candidate passed, NIMS staff is alerted and an official certificate will be mailed to the address listed on the candidate's profile. If the candidate failed, he or she may retake the exam any time. The exam fee will be reassessed for each retake.

If a power outage, computer failure, or any other issue arises, which results in a pause in testing, the Online Testing Center will save the candidate's progress and the time that has lapsed. To continue an exam, the candidate will simply log in, click **Take Test**, have his or her proctor enter a proctor code, and continue from where he or she left off.

The Online Theory Exam Process

1. Once the sponsor and proctor have arranged a time and place for candidates to test (under proctor supervision), the proctor will ensure that one computer is available per candidate.

2. The candidate will then go to the NIMS Web site and log in with the username and password that were assigned following registration.

3. If a candidate cannot recall a username or password, contact NIMS Staff at (703)352-4971 or at support@nims-skills.org.

4. Have the candidate click the **Skip** button on the **Edit Profile** page to go to the **Welcome Screen**.

5. Have the candidate click **Purchase Test** on the menu to the left and then click the **Order** box for the particular exam he or she is planning to take. Next have the candidate click **Add to Cart**.

6. Have the candidate provide payment for the exam by typing in credit card information or by typing in a prearranged account code and click **Submit Payment**. *Accredited programs: The candidate's discounted exam fee will be reflected on the next page.*

7. On the **Order Confirmation** page, scroll to the bottom and click **Done**.

8. The next page will show all exams that have been purchased for this candidate. Click the **Take Test** button next to the particular test that is being taken that day.

9. The page will shift down and the **Test ID** will be automatically copied into the corresponding box. At this point the proctor will enter a proctor code into the second box and then click **Start Test**.

10. From here the candidate can simply follow the on-screen prompts and begin the Online Theory Exam.

1.4 UTILIZING THE PRACTICE TESTS

The practice tests in this guide were methodically developed to align to standards assessed by the NIMS Level I certification exams. Each theory test corresponds to the NIMS exam with the same title. After completing classroom and lab instruction on the tasks, skills, and knowledge assessed by a specific NIMS certification, use this study guide as a tool to identify your strengths and weaknesses prior to taking the NIMS exam.

The theory exams for each certification will contain a minimum of 50 questions. Chapters 8, CNC Milling Certifications, and 9, CNC Turning Certifications, have 70 questions. Chapters 8 and 9 are each divided into two sections. The first 40 questions apply to both the CNC Operator certification and the Programming, Setup, and Operations certification. Questions 40 through 70 apply only to the Programming, Setup, and Operations certification. If preparing for just the CNC Milling Operator or CNC Turning Operator certifications, complete only questions 1 through 40 in the respective chapters. Use the answer sheets provided in the back of this book to record your answers. As with the NIMS certification exams, you may use a pencil, paper, calculator, and *The Machinery's Handbook* or *Shop Reference for Students and Apprentices* while taking these practice exams. After completing the practice exam, use the answer key to check for correct and incorrect responses. Explanations for each question are provided in the section following the answer key. Analyze the explanation for each incorrect response. Use this information as a guide to understand this concept and acquire that knowledge prior to taking the NIMS certification exam.

1.5 TEST-TAKING STRATEGIES

There are many factors that influence our ability to successfully take exams. The following proven test-taking strategies can assist in helping those taking an important exam achieve maximum results. Practicing these recommended guidelines while completing these practice exams will help you apply them when taking the NIMS certification exams.

Preparing to take the exam:

- Concentrate study efforts on areas of known weakness.
- Study in multiple, short blocks of time rather than in one marathon session.
- Study in an environment free of distractions.

Prior to taking the exam:

- Come well rested.
- Eat a balanced meal before the exam.

- Bring the allowable resources to the exam site.
- Arrive at the exam site early.

While taking the exam:

- Manage your time effectively—spend a minute at the start to check the total number of questions and the time allotment. Divide the time in minutes by the number of questions, so you know roughly how many minutes you can spend on each question.

- Thoroughly read the entire question and each answer twice before answering.

- If you are unsure of a question, leave it blank and return to it at the end—write the question number on a piece of paper so you know exactly what questions you need to return to. You may find cues to the question(s) you skipped embedded in other questions.

- After finishing, return to any unanswered questions.

- Do not leave any blanks—your intuition may be right.

- Spend any remaining time to double-check your answers.

Measurement, Materials, and Safety Certification

2

2.1 PRACTICE THEORY EXAM—MEASUREMENT, MATERIALS, AND SAFETY

1. Lubrication specifications and intervals for machine tools and equipment can be found in which of the following?

   A. Preventative Maintenance Manual
   B. Right to Know Information Station
   C. The Machinery's Handbook
   D. Shop Safety Manual

2. The process of joining gage blocks together by forcing the air out from between the two surfaces and creating a vacuum strong enough to hold the gage block build together describes:

   A. bonding.
   B. lapping.
   C. wringing.
   D. None of the above

3. The quantity of parts to be inspected from a production run during a specified time period is stated in the company's:

   A. sampling plan.
   B. inspection plan.
   C. production plan.
   D. process plan.

4. A machinist uses the specifications pictured in the following image to produce a part. Dimension "B" measures 0.975″. This measurement designates this part as:

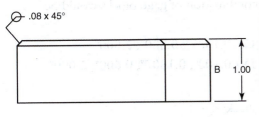

1	BLOCK	1.00 x 2.00 x 3.1	ALUM OR MILD STEEL
ITEM	DESCRIPTION	SIZE	MATERIAL

UNLESS OTHERWISE SPECIFIED DIMENSIONS ARE IN INCHES INTERPRET DIMENSIONS AND TOLERANCES PER ASME Y14.5M–1994	BENCHWORK – PERFORMANCE ASSESSMENT		
TOLERANCES .X ±.032 .XXX ±.005 .XX ±.015 ANGLES ± 1 DEG. FRACTIONS ± 1/64	DESIGNER		MATERIAL ALUMINUM OR MILD STEEL
	DWG CHK		
	DWG APPD		
	SCALE: NTS	DWG.:	SHEET 1 OF 2

   A. rework.
   B. a reject.
   C. a setup piece.
   D. out of control.

5. A machinist has finished using a workstation and is starting to clean up the metal chips produced. Which method listed here should be used to perform this cleanup safely?

   A. Wear a face shield and use compressed air.
   B. Use a chip brush.
   C. Push the chips onto the floor with your hand.
   D. Leave the chips on the machine until the end-of-week cleaning.

6. Using a grinding wheel that is too _____ on a surface grinding operation can produce excessive _____, which may lead to a bowed, warped, or twisted workpiece.

    A. hard; heat

    B. soft; RPM

    C. wide; chatter

    D. big in diameter; runout

7. A machinist is having difficulties with the tapped holes while assembling a fixture comprised of several machined components. When inserting the bolts into the 3/8-16 UNC 2B holes, the threads are very tight. After verifying the commercial bolts are within spec, which of the following answers could be the cause?

    A. The tap is worn.

    B. The tap drill cuts oversize.

    C. The tap drill diameter is too large.

    D. A bottoming tap is used.

8. What is the measurement in the following image?

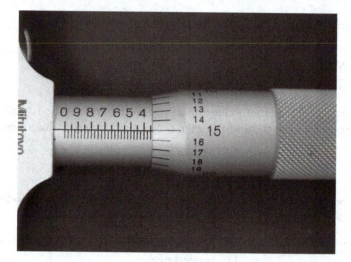

    A. 0.465″

    B. 0.340″

    C. 0.315″

    D. 0.365″

9. Using a standard 81-piece gage block set, which combination of gage blocks would be selected to make a gage block build of 2.8337″?

    A. 0.1007″, 0.033″, 0.100″, 0.600″, 2.000″

    B. 0.1007″, 0.133″, 0.600″, 2.000″

    C. 0.1337″, 0.800″, 2.000″

    D. 0.103″, 0.1307″, 0.600″, 2.000″

10. A comparison gage and profilometer are used to check:

    A. luminescence.

    B. thread percentage.

    C. radii.

    D. surface finish.

11. Which of the following is **not** an adequate PPE for a machining environment with an overhead crane and 115-dB sound levels?

    A. ANSI Z87.1-approved eyewear

    B. Industrial reusable ear plugs

    C. Neoprene (rubber) gloves for parts washing

    D. Prescription eyewear when walking across the shop floor

    E. Approved hard hat

12. A type of fitting that is used to apply grease with a grease gun to a lubrication point on a machine or a machine's component is called a:

    A. Zerk fitting.   C. ball oiler.
    B. quick connect fitting.   D. grease tip.

13. Which of the following symbols specifies a counterbore?

    A. ⊔   C. ↓
    B. ∨   D. ◎

14. Contains no iron, is nonmagnetic, aluminum all identify:

    A. nonferrous metal.   C. stainless steel.
    B. ferrous metal.   D. super alloy.

15. Prior to _____ a dimension, the _____ must be analyzed to determine what measuring tool should be used.

    A. inspecting; tolerances   C. reworking; part print
    B. inspecting; process plan   D. cutting; sampling plan

16. What is the measurement in the following image?

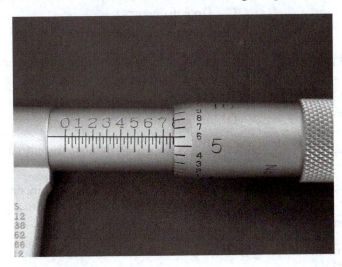

    A. 0.781″   C. 0.756″
    B. 0.806″   D. 0.731″

17. As a twist drill breaks through the bottom of a workpiece, the jagged edges of the partially finished hole could cause which of the following to occur?

    A. The machinist could reduce feed pressure and then finish drilling through.   C. The twist drill could bind or suddenly become jammed and break.
    B. The twist drill could bind or suddenly become jammed and slip in the chuck.   D. All of the above could occur.
                                                                                        E. Only B and C could occur.

18. A machinist is planning a machining operation and wants to prevent the high-speed steel (HSS) endmill from wearing prematurely. Which preventative measure will help extend the life of this tool?

    A. Increase the depth of cut by 20%.

    B. Use a collet to hold the tool.

    C. Increase the spindle RPM.

    D. Decrease the spindle RPM.

19. Created in 1971, and part of the U.S. Department of Labor, which agency enforces occupational safety standards and guides employers with training to protect their workforces?

    A. Occupational Safety and Health Administration

    B. National Institutes of Health

    C. Code of Federal Regulators

    D. Bureau of Labor Statistics

20. Which of the following is an applicable use for gage blocks?

    A. To measure the width of a slot

    B. As a standard to calibrate a micrometer

    C. To set an angle on a sine plate

    D. All of the above

    E. Only A and B

21. What geometric feature listed below would an adjustable parallel and a micrometer be an applicable method for inspection?

    A. Slot depth

    B. Slot width

    C. Hole diameter

    D. Countersink diameter

22. A reservoir with a built-in hand pump used to distribute oil to a machine's lubrication points is called a(n):

    A. oil shooter.

    B. reservoir pump.

    C. one-shot.

    D. oil gun.

23. The symbol $125\sqrt{}$ indicates _____ and is _____ in comparison to a $250\sqrt{}$.

    A. the surface finish of 125 microinches or better is required; better

    B. the carbon percent is 25%; softer

    C. the surface finish of 125 microinches or better is required; worse

    D. the carbon percent is 25%; harder

24. Which of the following could cause excessive heat and a warped workpiece after a surface grinding operation is performed?

    A. Depth of cut is excessive.

    B. The wheel is too hard.

    C. The wheel surface is dressed too fine.

    D. All of the above.

    E. None of the above.

25. Which of the following would **not** cause a reamer to cut oversize?

    A. Misalignment between the reamer and the drilled hole

    B. An RPM that is too fast

    C. Excessive material being left in the hole for the reaming operation

    D. An RPM that is too slow

26. The metal chips produced during machining operations should be treated as a:

    A. machining by-product.

    B. loss of profit.

    C. safety hazard.

    D. a consequence of the job.

27. Which technique listed here is **not** safe for lifting heavy object?

    A. Focus eyes straight ahead.

    B. Keep back straight.

    C. Position feet together.

    D. Bend knees.

    E. All of the above are safe.

28. A QC inspector randomly selects parts produced during a shift on a CNC turning center. The parts are inspected according to the procedures detailed in the inspection plan. What is the term given to the parts that were inspected?

    A. Range

    B. Critical parts

    C. Sampling

    D. Specimen

29. What measurement is closest to the following image?

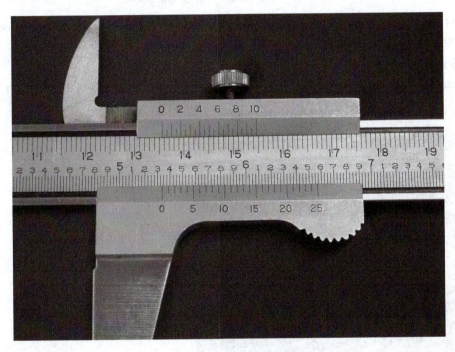

    A. 5.341″

    B. 4.825″

    C. 6.575″

    D. 5.325″

30. For a fire to burn, three components are necessary. Which of the following choices identifies the three components?

    A. Friction, air, heat

    B. CO_2, H_2O, a flame

    C. Wood, water, wind

    D. Heat, fuel, oxygen

31. A machinist is cleaning metal chips from a machine with an air hose when a coworker walks past and a particle becomes lodged in his eye. The first action should be:

    A. pull the top eyelid over the bottom eyelid.

    B. put the air hose away before the supervisor sees it.

    C. rub the eye in the clockwise direction.

    D. use a light stream of compressed air to blow the particle out.

32. In which of the following situations is gloves **not** acceptable PPE?

   A. Wearing canvas gloves to hold a hot part while working on a pedestal grinder

   B. Wearing neoprene (rubber) gloves to pour a chemical

   C. Wearing latex gloves while helping a coworker with a cut finger

   D. Wearing leather gloves to handle steel bar stock

33. A safe work area would include which of the following?

   A. A floor that has been swept and is free of oil spills

   B. Machines with all physical guards installed

   C. Raw materials stored on racks

   D. Hand and measuring tools cleaned and put away when not in use

   E. All of the above

34. Which OSHA standard utilizes locks and tags to isolate power and disable equipment while machine tools are being repaired?

   A. Energy isolation program

   B. Maintenance worker protection system

   C. Lockout/tagout

   D. Hazardous machine isolation system

35. What is the measurement in the following image?

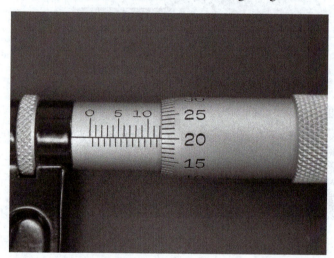

   A. 12.70 mm

   B. 10.22 mm

   C. 12.20 mm

   D. 11.70 mm

36. Which scenario described here would be considered the most unsafe?

   A. A machinist is using an electric hand drill with a frayed cord.

   B. The handle of a ball-peen hammer is cracked.

   C. A coworker is observed using a screwdriver as a chisel.

   D. The new apprentice is hammering on the end of a combination wrench in an effort to loosen a bolt.

   E. All of the above are unsafe and should be corrected immediately.

37. Select the correct gage block build for setting an 11° angle on a 5.0″ sine plate.

   A. 4.908″

   B. 0.9719″

   C. 0.9540″

   D. 4.8092″

38. Cutting internal threads with a worn tap could result in _____ threads due to a reduction in the _____ diameter:

    A. tight; minor
    B. tight; major
    C. loose; minor
    D. loose; major

39. An additive-enhanced, oil-based cutting fluid with excellent lubricity and good performance on drilling and tapping operations is:

    A. water-soluble oil.
    B. high-viscosity oil.
    C. mineral oil.
    D. sulfurized cutting oil.

40. 1070 Plain Carbon Steel contains _____ carbon than 1020 Plain Carbon Steel:

    A. an equal amount
    B. more
    C. less
    D. The answer cannot be determined from the information provided.

41. Material becoming embedded in the teeth of a file refers to _____, and if not cleaned, this could lead to poor cutting and scratches on the workpiece:

    A. pinning
    B. packing
    C. glazing
    D. None of the above

42. A manufacturing company is switching the type of coolant currently used in its CNC Turning Department. Information about the reactivity of the new coolant can be found on the _____, and employees should be formally trained about how to protect themselves when handling this coolant at the regularly scheduled _____.

    A. container label; production meeting
    B. inventory control sheet; awards luncheon
    C. PPE; department meeting
    D. MSDS; safety meeting

43. Which method listed here is considered safe and acceptable?

    A. Using a file to pry two fixture components apart
    B. Pushing a wrench away from you to loosen a bolt
    C. Carrying a scriber in your pocket
    D. Establishing a balanced position and pulling a wrench toward you to loosen a bolt

44. A _____ is a tapered wedge that can be adjusted to reduce the side play, which can develop over time from use, on the sliding components of a machine tool.

    A. drift
    B. gib
    C. way
    D. shim

45. A part print specifies a part's thickness at $1.000'' \pm 0.003''$ and a quantity of 300 parts. Which of the following inspection methods is both accurate and the most efficient to inspect this dimension during the machining process?

    A. Use a 0–1″ micrometer and a 1–2″ micrometer.
    B. Use a 6″ dial caliper.
    C. Use a height gage with a surface plate.
    D. Use a dial indicator mounted on a surface gage, calibrated to 1.000″ gage block, with a surface plate.

46. Which choice listed here correctly identifies the files shown in the following figure?

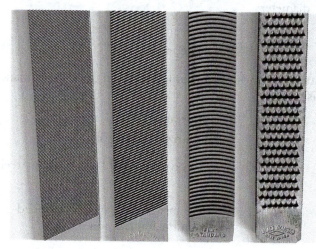

A. Mill, square, flat, half round

B. Bimetal, diamond, alternate, wavy

C. Jeweler, diemaker, rough cut, hook

D. Single cut, double cut, curved tooth, rasp tooth

47. A 5/16″ tap drill is used to create a hole prior to tapping a 3/8-UNC thread. Using the *Shop Reference for Students and Apprentices* or *The Machinery's Handbook* as a reference, find the percentage of thread that will be achieved.

A. 65

B. 70

C. 77

D. 81

48. A class of fit suitable for steel parts, or for shrink fits on light sections, and is considered the tightest fit permissible on external cast iron features describes:

A. FN 2 medium drive fit.

B. RC 2 sliding fit.

C. RC 4 close running fit.

D. FN 3 heavy drive fit.

49. What is the hole tolerance for a 2.250″ RC 2 sliding fit?

A. 0.000″ to 0.0007″

B. 0.000″ to 0.007″

C. 0.0004″ to 0.009″

D. 0.000″ to 0.007″

50. A part print specifies a slot to be machined on the workpiece at 0.875″ ± 0.003″ wide. The depth of the slot is specified at 0.375″ ± 0.003″. Which method would be the best choice for machining this slot?

A. One pass with a 4 flute, 7/8″ Ø endmill at 3/8″ depth

B. Two passes with a 0.4375″ Ø endmill at 3/8″ depth

C. One roughing pass with a 3/4″ Ø endmill at 0.360″ depth and then multiple finishing passes with a 0.500″ Ø endmill at 0.375″ depth

D. Four passes with a 7/32″ Ø endmill at 0.375″ depth

2.2 ANSWER KEY

1. A
2. C
3. A
4. B
5. B
6. A
7. A
8. B
9. B
10. D
11. D
12. A
13. A
14. A
15. A
16. A
17. D
18. D
19. A
20. D
21. B
22. C
23. A
24. D
25. D

26. C
27. C
28. C
29. A
30. D
31. A
32. A
33. E
34. C
35. C
36. E
37. C
38. B
39. D
40. B
41. A
42. D
43. D
44. B
45. D
46. D
47. C
48. A
49. A
50. C

2.3 PRACTICE THEORY EXAM EXPLANATIONS

1. Lubrication specifications and intervals for machine tools and equipment can be found in which of the following?

   A. Preventative Maintenance Manual   C. *The Machinery's Handbook*

   B. Right to Know Information Station   D. Shop Safety Manual

   Answer A is correct. Often, companies with multiple machine tools and other manufacturing-related equipment will develop and utilize a Preventative Maintenance Manual to ensure maintenance tasks are completed at the intervals and per the specifications recommended by the equipment manufacturers.

2. The process of joining gage blocks together by forcing the air out from between the two surfaces and creating a vacuum strong enough to hold the gage block build together describes:

   A. bonding.   B. lapping.

   C. wringing.   D. None of the above

   Answer C is correct. Wringing gage blocks together allows machinists and inspectors to create a gage block build of almost any size.

3. The quantity of parts to be inspected from a production run during a specified time period is stated in the company's:

   A. sampling plan.   B. inspection plan.

   C. production plan.   D. process plan.

   Answer A is correct. A sampling plan is a written procedure that includes the frequency of part inspections and the number of parts in a lot that shall be inspected.

4. A machinist uses the specifications pictured in the following image to produce a part. Dimension "B" measures 0.975″. This measurement designates this part as:

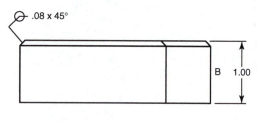

   A. rework.   B. a reject.

   C. a setup piece.   D. out of control.

   Answer B is correct. A reject is any part with one or more dimensions that are not within the tolerances specified on the part print.

5. A machinist has finished using a workstation and is starting to clean up the metal chips produced. Which method listed here should be used to perform this cleanup safely?

    A. Wear a face shield and use compressed air.

    B. Use a chip brush.

    C. Push the chips onto the floor with your hand.

    D. Leave the chips on the machine until the end-of-week cleaning.

    Answer B is correct. Using a chip brush is a safe method to remove metal chips. Using a brush prevents metal splinters from embedding in the hands or being cut by incidental contact with the chips.

6. Using a grinding wheel that is too _____ on a surface grinding operation can produce excessive _____, which may lead to a bowed, warped, or twisted workpiece.

    A. hard; heat

    B. soft; RPM

    C. wide; chatter

    D. big in diameter; runout

    Answer A is correct. A grinding wheel that is too hard for the material being ground will result in the grains of the wheel not releasing during the grinding operation. This will lead to the grains on the surface of the wheel becoming dull, which will cause wheel glazing, resulting in excessive heat buildup.

7. A machinist is having difficulties with the tapped holes while assembling a fixture comprised of several machined components. When inserting the bolts into the 3/8-16 UNC 2B holes, the threads are very tight. After verifying the commercial bolts are within spec, which of the following answers could be the cause?

    A. The tap is worn.

    B. The tap drill cuts oversize.

    C. The tap drill diameter is too large.

    D. A bottoming tap is used.

    Answer A is correct. As a tap wears, its cutting edges will diminish, resulting in reduced depth of cut and slightly smaller thread geometry. This combined with inefficient cutting action will result in greater material interference between the bolt and the internal thread.

8. What is the measurement in the following image?

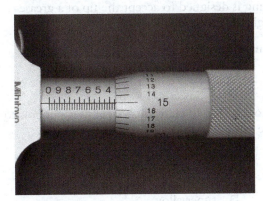

    A. 0.465″

    B. 0.340″

    C. 0.315″

    D. 0.365″

    Answer B is correct. The 3 on the sleeve of the depth micrometer is not visible, indicating 0.300″ is added to the reading. The first vertical line on the sleeve is also not visible, indicating 0.025″ is added to the reading. The horizontal line numbered 15 on the thimble aligns to the horizontal line running the length of the sleeve, indicating 0.015″ is added to the reading. 0.300″ + 0.025″ + 0.015″ = 0.340″

9. Using a standard 81-piece gage block set, which combination of gage blocks would be selected to make a gage block build of 2.8337"?

   A. 0.1007", 0.033", 0.100", 0.600", 2.000"

   B. 0.1007", 0.133", 0.600", 2.000"

   C. 0.1337", 0.800", 2.000"

   D. 0.103", 0.1307", 0.600", 2.000"

   Answer B is correct. The gage blocks 0.1007", 0.133", 0.600", 2.000" are all part of a standard 81-piece set and when combined equal 2.83337".

10. A comparison gage and profilometer are used to check:

   A. luminescence.

   B. thread percentage.

   C. radii.

   D. surface finish.

   Answer D is correct. The roughness and waviness of a part's surface are characteristics that can be inspected by comparing the surface to samples on a comparison gage or checking with an electronic device called a profilometer.

11. Which of the following is not an adequate PPE for a machining environment with an overhead crane and 115-dB sound levels?

   A. ANSI Z87.1-approved eyewear

   B. industrial reusable ear plugs

   C. neoprene (rubber) gloves for parts washing

   D. prescription eyewear when walking across the shop floor

   E. approved hard hat

   Answer D is correct. Everyday prescription glasses are not designed to withstand much more than a minimal impact and are not an acceptable PPE. Furthermore, the broken glass of a regular prescription lens can injure the eye.

12. A type of fitting that is used to apply grease with a grease gun to a lubrication point on a machine or a machine's component is called a:

   A. Zerk fitting.

   B. quick connect fitting.

   C. ball oiler.

   D. grease tip.

   Answer A is correct. The tip of the Zerk fitting is designed to accept the tip of a grease gun. The grease gun pumps the grease through the fitting.

13. Which of the following symbols specifies a counterbore?

   A. ⊔

   B. ∨

   C. ↓

   D. ◎

   Answer A is correct. The geometric dimensioning and tolerancing system uses symbols to specify geometric features on a part print. The symbol indicating a counterbore is ⊔.

14. Contains no iron, is nonmagnetic, aluminum all identify:

   A. nonferrous metal.

   B. ferrous metal.

   C. stainless steel.

   D. super alloy.

   Answer A is correct. Nonferrous metals are those that do not contain iron and are therefore nonmagnetic. Aluminum is one example of a nonferrous metal.

15. Prior to _____ a dimension, the _____ must be analyzed to determine what measuring tool should be used.

   A. inspecting; tolerances   C. reworking; part print
   B. inspecting; process plan   D. cutting; sampling plan

   Answer A is correct. The tolerance specifies the size range that is acceptable for a dimension. Prior to inspecting a dimension the tolerance should be studied so the most appropriate
   measuring tool can be selected.

16. What is the measurement in the following image?

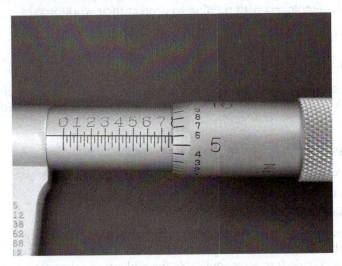

   A. 0.781″   C. 0.756″
   B. 0.806″   D. 0.731″

   Answer A is correct. The beveled edge of the thimble is past the 7 on the sleeve, indicating
   0.700″ is added to the measurement. The beveled edge of the thimble is past the third vertical line on the sleeve, indicating 0.075″ is added to the measurement. The horizontal line
   numbered 6 on the thimble aligns to the horizontal line running the length of the sleeve,
   indicating 0.006″ is added to the reading. 0.700 + 0.075″ + 0.006″ = 0.781″

17. As a twist drill breaks through the bottom of a workpiece, the jagged edges of the partially
finished hole could cause which of the following to occur?

   A. The machinist could reduce feed pres-   C. The twist drill could bind or suddenly
       sure and then finish drilling through.   become jammed and break.
   B. The twist drill could bind or suddenly   D. All of the above could occur.
       become jammed and slip in the chuck.   E. Only B and C could occur.

   Answer D is correct. As a twist drill approaches the "breakthrough" point of a hole, the
   drill bit may "grab." This could result in the drill bit slipping in the chuck or it could become jammed and break. Reducing the feed pressure is one strategy a machinist can use to
   prevent the twist drill from slipping and breaking.

18. A machinist is planning a machining operation and wants to prevent the HSS endmill from wearing prematurely. Which preventative measure will help extend the life of this tool?

    A. Increase the depth of cut by 20%.   C. Increase the spindle RPM.
    B. Use a collet to hold the tool.   D. Decrease the spindle RPM.

    Answer D is correct. A spindle speed that is too fast will lead to excessive heat or wear to the cutting tool. This accelerated wear could result in damage to the tool and workpiece during the machining operation.

19. Created in 1971, and part of the U.S. Department of Labor, which agency enforces occupational safety standards and guides employers with training to protect their workforces?

    A. Occupational Safety and Health   C. Code of Federal Regulators
       Administration   D. Bureau of Labor Statistics
    B. National Institutes of Health

    Answer A is correct. The Occupational Safety and Health Administration (OSHA) was created to assure the safety of working men and women by developing and enforcing standards and educating the workforce.

20. Which of the following is an applicable use for gage blocks?

    A. To measure the width of a slot   D. All of the above
    B. As a standard to calibrate a micrometer E. Only A and B
    C. To set an angle on a sine plate

    Answer D is correct. Gage block sets are manufactured in different grades of accuracy. Grades AS-1 and AS-2 are suitable for general shop applications such as slot measurement and setting angles on a sine plate. Grade 0 sets are applicable for calibrating other measuring tools.

21. Which geometric feature listed here would an adjustable parallel and a micrometer be an applicable method for inspection?

    A. Slot depth   C. Hole diameter
    B. Slot width   D. Countersink diameter

    Answer B is correct. An adjustable parallel can be adjusted against the surfaces of a slot and then tightened with a locking screw. The parallel can then be removed, and a measurement can be obtained by measuring the width with an outside micrometer.

22. A reservoir with a built-in hand pump used to distribute oil to a machine's lubrication points is called a(n):

    A. oil shooter.   C. one-shot.
    B. reservoir pump.   D. oil gun.

    Answer C is correct. A "one-shot" uses the pressure created by its hand pump to distribute oil to the machine's lubrication points.

23. The symbol $125\sqrt{}$ indicates _____ and is _____ in comparison to a $250\sqrt{}$:

    A. the surface finish of 125 microinches or better is required; better

    C. the surface finish of 125 microinches or better is required; worse

    B. the carbon percent is 25%; softer

    D. the carbon percent is 25%; harder

    Answer A is correct. Surface finish is measured in microinches, with measurements ranging from 0.5 to 1000. The smaller the value of a surface finish measurement, the smoother the surface will be.

24. Which of the following could cause excessive heat and a warped workpiece after a surface grinding operation is performed?

    A. Depth of cut is excessive.

    D. All of the above.

    B. The wheel is too hard.

    E. None of the above.

    C. The wheel surface is dressed too fine.

    Answer D is correct. The heat generated by a surface grinding operation can be minimized if the appropriate techniques and methods are used. Conversely, not following the recommended techniques and methods will lead to overheating and a warped or distorted workpiece. The depth of cut, hardness of the wheel, and methods used to dress the grinding wheel are variables that can influence overheating.

25. Which of the following would **not** cause a reamer to cut oversize?

    A. Misalignment between the reamer and the drilled hole

    C. Excessive material being left in the hole for the reaming operation

    B. An RPM that is too fast

    D. An RPM that is too slow

    Answer D is correct. There are several variables that will cause a reamer to produce an oversize hole. An RPM that is too slow will result in the inefficiency of the reamer's cutting action but will not lead to an oversize reamed hole.

26. The metal chips produced during machining operations should be treated as a:

    A. machining by-product.

    C. safety hazard.

    B. loss of profit.

    D. a consequence of the job.

    Answer C is correct. Metal chips have the potential to cause injury and should be handled as potentially hazardous.

27. Which technique listed here is **not** safe for lifting heavy object?

    A. Focus eyes straight ahead.

    D. Bend knees.

    B. Keep back straight.

    E. All of the above are safe.

    C. Position feet together.

    Answer C is correct. When lifting heavy objects, the feet should be positioned shoulder-width apart to establish good balance.

28. A QC inspector randomly selects parts produced during a shift on a CNC turning center. The parts are inspected according to the procedures detailed in the inspection plan. What is the term given to the parts that were inspected?

A. Range

B. Critical parts

C. Sampling

D. Specimen

Answer C is correct. A sampling is a portion of the whole (population) lot of parts that when inspected is representative of that larger population.

29. What measurement is closest to the following image?

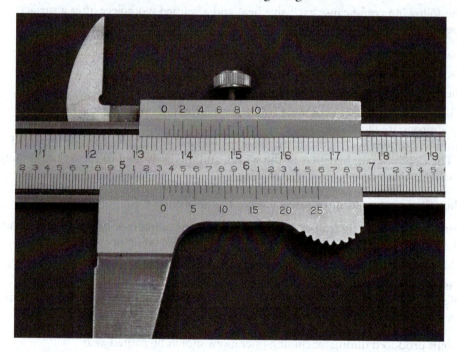

A. 5.341″

B. 4.825″

C. 6.575″

D. 5.325″

Answer A is correct. The vertical 0 line on the sliding jaw is past the whole number 5 on the main scale, indicating 5.00″ is added to the measurement. The vertical 0 line on the sliding jaw is also past the 3 on the smaller graduations on the main scale, indicating 0.300″ is added to the measurement. The vertical 0 line on the sliding jaw is also past one of the smaller unnumbered graduations on the main scale, indicating 0.025″ is added to the measurement. The 16 on the sliding vernier scale is aligned best with the line above it, indicating 0.016″ is added to the measurement. 5.00″ + 0.300″ + 0.025″ + 0.016″ = 5.341″

30. For a fire to burn, three components are necessary. Which of the following choices identifies the three components?

A. Friction, air, heat

B. CO_2, H_2O, a flame

C. Wood, water, wind

D. Heat, fuel, oxygen

Answer D is correct. The three sides of a fire triangle are heat, fuel, and oxygen. If one of these elements is removed, the fire will not burn.

31. A machinist is cleaning metal chips from a machine with an air hose when a coworker walks past and a particle becomes lodged in his eye. The first action should be:

    A. pull the top eyelid over the bottom eyelid.

    B. put the air hose away before the supervisor sees it.

    C. rub the eye in the clockwise direction.

    D. use a light stream of compressed air to blow the particle out.

    Answer A is correct. The action of pulling the upper eyelid over the lower eyelid could release the particle.

32. In which of the following situations is gloves **not** acceptable PPE?

    A. Wearing canvas gloves to hold a hot part while working on a pedestal grinder

    B. Wearing neoprene (rubber) gloves to pour a chemical

    C. Wearing latex gloves while helping a coworker with a cut finger

    D. Wearing leather gloves to handle steel bar stock

    Answer A is correct. Wearing gloves while working on a pedestal grinder could lead to the glove getting pulled into the grinding wheel and injury to the hand or fingers.

33. A safe work area would include which of the following?

    A. A floor that has been swept and is free of oil spills

    B. Machines with all physical guards installed

    C. Raw materials stored on racks

    D. Hand and measuring tools cleaned and put away when not in use

    E. All of the above

    Answer E is correct. Good housekeeping practices will aid in keeping the work area organized and preventing injuries. Machines should have all physical guards properly installed before they are operated.

34. Which OSHA standard utilizes locks and tags to isolate power and disable equipment while machine tools are being repaired?

    A. Energy isolation program

    B. Maintenance worker protection system

    C. Lockout/tagout

    D. Hazardous machine isolation system

    Answer C is correct. Lockout/tagout is the OSHA standard that requires the use of locks and tags to isolate the power to equipment when repairs are being performed.

35. What is the measurement in the following image?

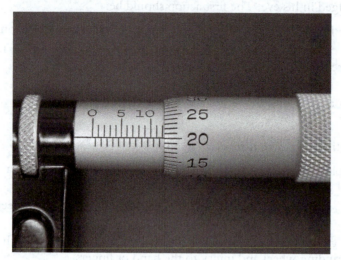

A. 12.70 mm
B. 10.22 mm
C. 12.20 mm
D. 11.70 mm

Answer C is correct. The beveled edge of the thimble is past the 12 line on the sleeve, indicating 12 mm is added to the measurement. The horizontal line numbered 20 on the thimble aligns with the horizontal line running the length of the sleeve, indicating .20 mm is added to the measurement. 12 mm + 0.20 mm = 12.20 mm

36. Which scenario described here would be considered the most unsafe?

A. A machinist is using an electric hand drill with a frayed cord.

B. The handle of a ball-peen hammer is cracked.

C. A coworker is observed using a screwdriver as a chisel.

D. The new apprentice is hammering on the end of a combination wrench in an effort to loosen a bolt.

E. All of the above are unsafe and should be corrected immediately.

Answer E is correct. To work safely and prevent injuries, tools must be used for their intended purpose and they must be inspected for damage prior to using them.

37. Select the correct gage block build for setting an 11° angle on a 5.0″ sine plate.

A. 4.908″
B. 0.9719″
C. 0.9540″
D. 4.8092″

Answer C is correct. The sine plate and gage block build form a right triangle. Using the sine function, the gage block build or opposite side of the triangle can be calculated.

Sine 11° = Opposite / 5.0″

Opposite = Sine 11° × 5.0″

Opposite = 0.9540″

38. Cutting internal threads with a worn tap could result in _____ threads due to a reduction in the _____ diameter:

A. tight; minor
B. tight; major
C. loose; minor
D. loose; major

Answer B is correct. The dull cutting edges of a worn tap will result in the tap geometry being smaller than the thread standard. Cutting threads with this tap will result in less material removal and greater material interference between the external and internal threads.

39. An additive-enhanced, oil-based cutting fluid with excellent lubricity and good performance on drilling and tapping operations is:

A. water-soluble oil.

C. mineral oil.

B. high-viscosity oil.

D. sulfurized cutting oil.

Answer D is correct. Using cutting oil with additives such as sulfur improves lubricity between the cutting tool and material. This works well on applications where there is considerable contact between the tool and material such as drilling and tapping.

40. 1070 Plain Carbon Steel contains _____ carbon than 1020 Plain Carbon Steel:

A. an equal amount

C. less

B. more

D. The answer cannot be determined from the information provided.

Answer B is correct. The last two digits of a steel designation under the AISI/SAE numbering system indicate the percentage of carbon the material contains. The carbon content is specified in hundredths of a percent. 1070 contains 0.70% carbon and 1020 contains 0.20% carbon.

41. Material becoming embedded in the teeth of a file refers to _____, and if not cleaned, this could lead to poor cutting and scratches on the workpiece:

A. pinning

C. glazing

B. packing

D. None of the above

Answer A is correct. The material being removed by a file is cut into particles called pins. The pins will collect in the teeth of the file, causing inefficient cutting conditions.

42. A manufacturing company is switching the type of coolant currently used in its CNC Turning Department. Information about the reactivity of the new coolant can be found on the _____, and employees should be formally trained about how to protect themselves when handling this coolant at the regularly scheduled _____.

A. container label; production meeting

C. PPE; department meeting

B. inventory control sheet; awards luncheon

D. MSDS; safety meeting

Answer D is correct. Material safety data sheets (MSDSs) identify the potential hazards associated with workplace materials. Training on handling these hazards can be conducted at regularly held safety meetings.

43. Which method listed here is considered safe and acceptable?

A. Using a file to pry two fixture components apart

C. Carrying a scriber in your pocket

B. Pushing a wrench away from you to loosen a bolt

D. Establishing a balanced position and pulling a wrench toward you to loosen a bolt

Answer D is correct. When loosening or tightening a threaded fastener with a wrench, pulling the wrench toward you will reduce the opportunity for injuries if the wrench slips or the fastener suddenly becomes loose.

44. A _____ is a tapered wedge that can be adjusted to reduce the side play, which can develop over time from use, on the sliding components of a machine tool.

A. drift

B. gib

C. way

D. shim

Answer B is correct. The dovetail-shaped slides of a machine tool use an adjustable gib that can be tightened to alleviate the wear that occurs from continuous use.

45. A part print specifies a part's thickness at 1.000″ ± 0.003″ and a quantity of 300 parts. Which of the following inspection methods is both accurate and the most efficient to inspect this dimension during the machining process?

A. Use a 0–1″ micrometer and a 1–2″ micrometer.

B. Use a 6″ dial caliper.

C. Use a height gage with a surface plate.

D. Use a dial indicator mounted on a surface gage, calibrated to 1.000″ gage block, with a surface plate.

Answer D is correct. "Zeroing" a dial indicator mounted on a surface gage to the height of the 1.0″ gage blocks will provide a reference. The manufactured parts can then easily be checked by sliding the parts under the indicator. As long as a part's variation to the "zeroed" indicator does not exceed the tolerance, it is a good part.

46. Which choice listed here correctly identifies the files shown in the following figure?

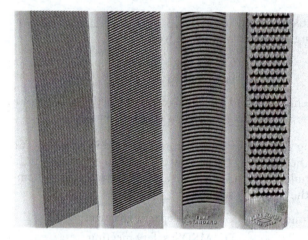

A. mill, square, flat, half round

B. bimetal, diamond, alternate, wavy

C. jeweler, diemaker, rough cut, hook

D. single cut, double cut, curved tooth, rasp tooth

Answer D is correct. The files show the different types of tooth configurations and are arranged as follows: single cut, double cut, curved tooth, and rasp tooth.

47. A 5/16″ tap drill is used to create a hole prior to tapping a 3/8-UNC thread. Using the *Shop Reference for Students and Apprentices* or *The Machinery's Handbook* as a reference, find the percentage of thread that will be achieved.

A. 65

B. 70

C. 77

D. 81

Answer C is correct. The chart indicates using the 5/16″-diameter tap drill will result in 77% thread depth.

48. A class of fit suitable for steel parts, or for shrink fits on light sections, and is considered the tightest fit permissible on external cast iron features describes:

    A. FN 2 medium drive fit.

    B. RC 2 sliding fit.

    C. RC 4 close running fit.

    D. FN 3 heavy drive fit.

    Answer A is correct. A class of fit suitable for steel parts, or for shrink fits on light sections, and is considered the tightest fit permissible on external cast iron features defines a FN2 Medium drive fit.

49. What is the hole tolerance for a 2.250″ RC 2 sliding fit?

    A. 0.000″ to 0.0007″

    B. 0.000″ to 0.007″

    C. 0.0004″ to 0.009″

    D. 0.000″ to 0.007″

    Answer A is correct. Referencing the American National Standard Running and Sliding Fit chart, the hole tolerance for an RC 2 sliding fit of this size is 0.000 to 0.0007″.

50. A part print specifies a slot to be machined on the workpiece at 0.875″ ± 0.003″ wide. The depth of the slot is specified at 0.375″ ± 0.003″. Which method would be the best choice for machining this slot?

    A. One pass with a 4 flute, 7/8″ Ø end-mill at 3/8″ depth

    B. Two passes with a 0.4375″ Ø endmill at 3/8″ depth

    C. One roughing pass with a 3/4″ Ø endmill at .360″ depth and then multiple finishing passes with a 0.500″ Ø endmill at 0.375″ depth

    D. Four passes with a 7/32″ Ø endmill at 0.375″ depth

    Answer C is correct. A roughing pass along the centerline of the slot with a 3/4″ Ø endmill will remove the majority of the material and leave 0.0625″ on each side of the slot width for finishing. The 0.500″ Ø endmill can then be used to make successive passes to machine the slot to the specified width, depth, and location.

3 Job Planning, Benchwork, and Layout Certification

3.1 PRACTICE THEORY EXAM —JOB PLANNING, BENCHWORK, AND LAYOUT

1. Using the formula $TPI = \dfrac{D - d}{l}$, calculate the taper per inch of the part in Figure 3.1.

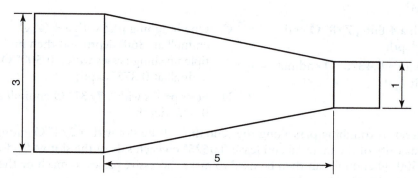

   A. 0.80″

   B. 1.25″

   C. 0.40″

   D. 0.45″

2. With two wheels pressed against the circumference of a workpiece, a(n) _____ tool forms a diamond or straight pattern.

   A. indexing

   B. turning

   C. knurling

   D. form turning

3. Which of the following is a semiprecision layout tool commonly used to produce angles up to 180°?

   A. Divider

   B. Protractor

   C. Parallel

   D. Angle plate

4. Using the formula $RPM = \dfrac{4 \cdot CS}{\varnothing}$, calculate the cutting speed for drilling a 7/8″ Ø hole at 90 SFPM.

   A. 393

   B. 322

   C. 411

   D. 205

5. A hole that does not go through a workpiece is called a _____ and requires the use of a _____ tap to produce threads to the full depth of the hole.

   A. blind hole; bottoming

   B. depth hole; plug

   C. stop hole; bottoming

   D. bottom hole; taper

6. Which of the following countersinks should be used to cut a countersink for an industrial flathead screw?

   A. 90°

   B. 60°

   C. 82°

   D. 100°

7. From top to bottom, the taps pictured here are arranged in what sequence?

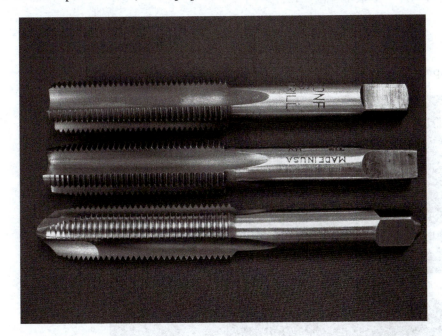

   A. Plug, taper, bottoming

   B. Taper, plug, bottoming

   C. Bottoming, plug, taper

   D. Bottoming, taper, plug

8. Prior to drilling a hole with a twist drill, a _____ is used to create an indentation and a _____ or _____ is used to produce a starter hole to prevent the twist drill from "walking off."

   A. drive punch; countersink; centerdrill

   B. center punch; centerdrill; spot drill

   C. press; spot face; tap drill

   D. hole punch; gun drill; countersink

9. The feed rate of a twist drill is specified in:

   A. TIR.

   B. RPM.

   C. IPM.

   D. IPR.

10. A tool equipped with either wire or short soft bristles and used to remove pins from a file is called a:

   A. pin remover.

   B. wire brush.

   C. file cleaner.

   D. file card.

11. Which drill would be the **best** choice to drill prior to reaming a 3/8″ Ø hole using a chucking reamer?

   A. 3/8″

   B. U

   C. 23/64″

   D. 5/16″

12. Line A-A in the following drawing is a _____ line.

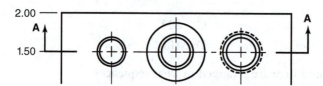

    A. leader

    B. hidden

    C. extension

    D. cutting plane

13. Select the semiprecision layout tool used for producing circles, radii, and arcs.

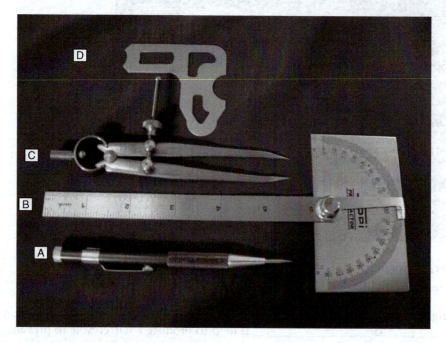

    A. Scriber

    B. Protractor

    C. Dividers

    D. Radius gage

14. The minor diameter and maximum pitch diameter for a 3/8-16 UNC-2A thread is:

    A. 0.299″; 0.333″.

    B. 0.321″; 0.334″.

    C. 0.3076″; 0.3376″.

    D. 0.324″; 0.347″.

15. Which of the following is a method of peripheral milling where the workpiece is fed with the rotation of the cutting tool?

    A. Conventional milling

    B. Face milling

    C. Precision milling

    D. Climb milling

16. Which angle is a solid square used to inspect?

    A. 45°

    B. 60°

    C. 80°

    D. 90°

17. What is the recommended keyseat depth for 5/8″ square key on the shaft detailed here?

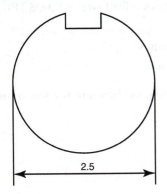

2.5

A. 0.290″   C. 0.600″

B. 0.375″   D. 0.313″

18. Which of the following statement(s) is(are) **true** about carbide cutting tools?

A. Carbide is more expensive than high-speed steel (HSS).

B. Carbide is brittle and easily chipped.

C. Carbide tools can perform at higher speeds and feeds than HSS.

D. All of the above are true.

19. When using a hacksaw, it is recommended to have at least _____ teeth in contact with the workpiece.

A. nine   C. five

B. seven   D. three

20. A machinist needs to cut 1/2″ off the end of a hardened steel rod (0.625 Ø × 7.5″ long). The tolerance is ± 1/16″. Select the most efficient process to complete this job.

A. Abrasive cutoff saw   C. Surface grinder

B. Wire EDM   D. Plasma torch

21. When cutting harder materials with an HSS endmill, tool life can be extended and tool failure can be prevented by:

A. increasing RPM.   C. doubling feed rates.

B. lowering cutting speeds.   D. multiplying the feed rate by 1.5.

22. _____ is commonly used in the layout process to enhance visibility of and to provide contrast to the layout lines.

A. Spray paint   C. Layout fluid

B. Buffing   D. Prussian blue

23. Which of the following is a result of draw filing?

A. Smooth surface finish   C. Slower rate of material removal

B. Fast rate of material removal   D. Both A and C

24. Using the formula $T = L/RPM \times$ feed rate, calculate the time to machine one pass of a 0.75″ Ø × 18.04″ long, 410 stainless steel shaft at 55 SFPM, with a feed rate of 0.003 IPR. Use 4 as the constant in the RPM formula.

    A. 18 minutes
    B. 20.5 minutes

    C. 36.4 minutes
    D. 54 minutes

25. What is the recommended tap drill for the threaded hole on the part shown in the following figure?

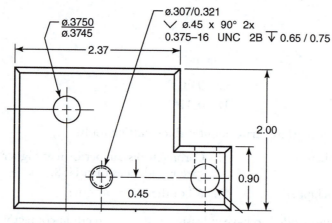

    A. 0.290″
    B. Q

    C. 3/8″
    D. 5/16″

26. Emory or silicon carbide abrasive cloth is most applicable for which of the following?

    A. Polishing scratches from metal surfaces
    B. Deburring
    C. Producing a smooth surface

    D. All of the above
    E. None of the above

27. What is the hole diameter for a 2″ Ø RC6 running fit? Use *The Machinery's Handbook* or *Shop Reference for Students and Apprentices* as a reference source.

    A. 1.9957″ to 1.9975″
    B. 1.997″ to 2.003″

    C. 2.000″ to 2.003″
    D. 2.000″ to 2.0003″

28. A _____ is a layout tool used to mark lines on metal surfaces. _____ is used to enhance the visibility of layout lines.

    A. square; A shop light
    B. center punch; A scriber

    C. scriber; Layout fluid
    D. scriber; A center punch

29. The total number of divisions on the thimble of an inch-based micrometer is:

    A. 100.
    B. 10.

    C. 15.
    D. 25.

30. A machinist is given a print with metric dimensions. The conversion of a 30-mm shaft dimension to the nearest 0.0001″ would be?

    A. 0.7620″
    B. 1.1811″

    C. 0.8467″
    D. 0.3000″

31. Refer to the following figure to determine how far past the 1.0″ depth it is necessary to drill, to compensate for the drill point (118° included angle drill; 0.375Ø).

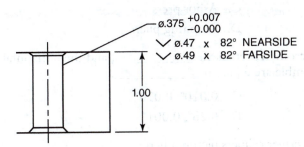

A. 0.1125″

B. 0.1875″

C. 0.2625″

D. 0.375″

32. The major alloy element of 2520 alloy steel is:

A. chromium.

B. tungsten.

C. manganese.

D. nickel.

33. Applying _____ to a file will aid in preventing material from becoming lodged in the teeth of the file.

A. cutting oil

B. chalk

C. WD-40

D. coolant

34. The stock allowance for a hand reaming operation is approximately:

A. 0.010″ to 0.020″.

B. 0.008″ to 0.015″.

C. 0.005″ to 0.010″.

D. 0.001″ to 0.008″.

35. Line A in the following print is a _____ line that indicates the specified geometry is _____ in the provided view.

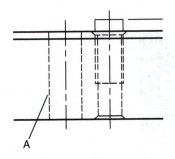

A. hidden; not visible

B. object; not visible

C. hidden; a cross section

D. phantom; visible

36. Which of the following statements **best** describes a center punch?

A. 60° point angle marks the intersections of arc and circle center points.

B. 60° point angle is used to enlarge a prick punch mark for drilling.

C. 90° point angle marks the intersections of arc and circle center points.

D. 90° point angle is used to enlarge a prick punch mark for drilling.

37. The machinist is assembling components that specify an FN2 medium drive fit. Which tool or tools should be used in this assembly process?

    A. Impact wrench
    B. Locking compound
    C. Arbor press
    D. Snap ring pliers

38. The vertical lines on the sleeve of a micrometer represent _____, and the horizontal lines around the circumference of a thimble are equal to _____.

    A. 0.025″; 0.001″
    B. 0.025″; 0.010″
    C. 0.010″; 0.010″
    D. 0.25″; 0.001″

39. What is the inch measurement on the vernier calipers pictured here?

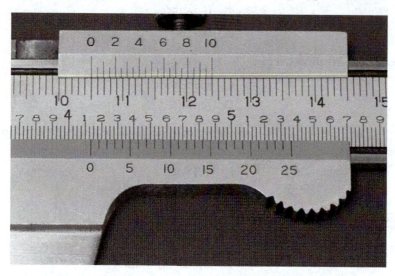

    A. 4.135″
    B. 5.360″
    C. 4.635″
    D. 4.110″

40. What is the measurement shown on the rule pictured here?

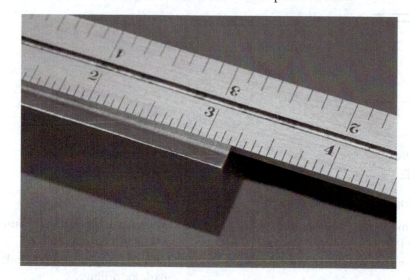

    A. 3 and 3/8″
    B. 3 and 3/4″
    C. 3 and 3/32″
    D. 3 and 3/16″

41. A vernier caliper can accurately measure to an increment as small as:

 A. 0.0010″.

 C. 0.0005″.

 B. 0.0001″.

 D. 0.0050″.

42. Which of the following tools should be used to finish machine hole A in the following drawing?

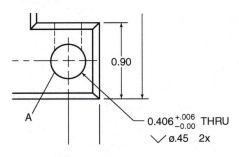

 A. Reamer

 C. Tap

 B. Twist drill

 D. Counterbore

43. What is the recommended stock allowance for machine reaming a hole between 1/4″ Ø and 1/2″ Ø?

 A. 0.015″

 C. 0.025″

 B. 0.0105″

 D. 0.020″

44. Which of the surface finish specifications are arranged in order from the smoothest to the roughest?

 A. $\overset{4}{\vee}$, 60 μin, $\overset{100}{\vee}$, $\overset{150}{\vee}$

 C. 225 μin, 125 μin, 50 μin, $\overset{25}{\vee}$

 B. 200 μin, 150 μin, $\overset{100}{\vee}$, $\overset{150}{\vee}$

 D. 4 μin, $\overset{40}{\vee}$, $\overset{400}{\vee}$, $\overset{150}{\vee}$

45. The RPM for reaming a 5/16″ Ø hole should be approximately _____ the RPM for drilling a hole of the same diameter?

 A. one-half

 C. equal

 B. double

 D. one-fourth

46. Calculate the feed rate using the following variables: 12 flute 2.5″ Ø face mill, 65 SFPM, and 0.006″ feed per tooth. Use 4 as the constant.

 A. 7.5 IPM

 C. 46.8 IPM

 B. 4.7 IPM

 D. 8.0 IPM

47. What is the minimum-diameter round stock needed to produce the wrench head specified on the following part?

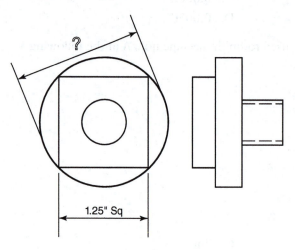

1.25" Sq

A. 1.266″   C. 1.5″

B. 1.7678″   D. 1.250″

48. Dimension A specifies a _____, and dimension B specifies a _____.

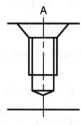 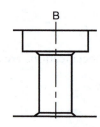

A. chamfer; spotface   C. countersink; counterbore

B. countersink; stepbore   D. counterbore; countersink

49. The distance a point on the circumference of a rotating cutting tool travels in 1 minute is expressed as:

A. revolutions per minute (RPM).   C. surface feet per minute (SFPM).

B. total indicator runout (TIR).   D. All of the above.

50. A flat spot machined on a rough or angled surface to provide a bearing surface for bolts, nuts, or washers describes a:

A. pilot.   C. counterbore.

B. spotface.   D. countersink.

3.2 ANSWER KEY

1.	C	**26.**	D
2.	C	**27.**	C
3.	B	**28.**	C
4.	C	**29.**	D
5.	A	**30.**	B
6.	C	**31.**	A
7.	C	**32.**	D
8.	B	**33.**	B
9.	D	**34.**	D
10.	D	**35.**	A
11.	C	**36.**	D
12.	D	**37.**	C
13.	C	**38.**	A
14.	A	**39.**	A
15.	D	**40.**	D
16.	D	**41.**	A
17.	D	**42.**	B
18.	D	**43.**	A
19.	D	**44.**	A
20.	A	**45.**	A
21.	B	**46.**	A
22.	C	**47.**	B
23.	D	**48.**	C
24.	B	**49.**	C
25.	D	**50.**	B

3.3 PRACTICE THEORY EXAM EXPLANATIONS

1. Using the formula TPI $= \dfrac{D - d}{l}$, calculate the taper per inch of the part in Figure 3.1.

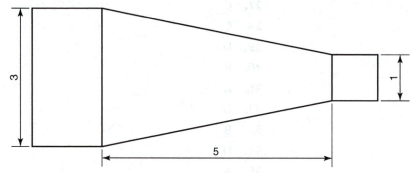

A. 0.80″

B. 1.25″

C. 0.40″

D. 0.45″

Answer C is correct. Using the formula TPI = 3 − 1/5. TPI = 2/5. TPI = 0.4″

2. With two wheels pressed against the circumference of a workpiece, a(n) _____ tool forms a diamond or straight pattern.

A. indexing

B. turning

C. knurling

D. form turning

Answer C is correct. A knurl is typically produced to provide grip to a cylindrical-shaped part, and it is produced by using a knurling tool on the lathe. A knurling tool has two wheels or rolls that when pressed against the workpiece will form the raised knurl pattern.

3. Which of the following is a semiprecision layout tool commonly used to produce angles up to 180°?

A. Divider

B. Protractor

C. Parallel

D. Angle plate

Answer B is correct. The plain protractor and the protractor head of a combination set are both semiprecision layout tools used to lay out angles.

4. Using the formula RPM $= \dfrac{4 \cdot CS}{\phi}$, calculate the cutting speed for drilling a 7/8″ Ø hole at 90 SFPM.

A. 393

B. 322

C. 411

D. 205

Answer C is correct. Inserting the provided variables into the RPM formula will result in RPM = 4 × 90/0.875; RPM = 360/0.875; RPM = 411.43.

5. A hole that does not go through a workpiece is called a _____ and requires the use of a _____ tap to produce threads to the full depth of the hole.

A. blind hole; bottoming

B. depth hole; plug

C. stop hole; bottoming

D. bottom hole; taper

Answer A is correct. A hole that is drilled partially through a workpiece to a depth is called a blind hole. A bottoming tap will have only a small chamfer on the end of the tap, which allows for it to cut full threads almost to the bottom of the blind hole.

6. Which of the following countersinks should be used to cut a countersink for an industrial flathead screw?

    A. 90°   C. 82°
    B. 60°   D. 100°

    Answer C is correct. A typical industrial flathead screw has a head with an 82° included angle. For this type of fastener to seat properly, the countersink will need to have the same angle.

7. From top to bottom, the taps pictured here are arranged in what sequence?

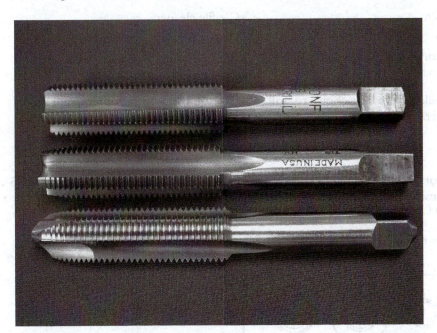

    A. Plug, taper, bottoming   C. Bottoming, plug, taper
    B. Taper, plug, bottoming   D. Bottoming, taper, plug

    Answer C is correct. The taps are identifiable by their chamfer types. The tap at the top has a minimal chamfer on only the first two threads and is a bottoming tap. The tap in the middle is a plug tap and has a chamfer that affects the first five threads. The tap at the bottom has a large lead-in chamfer that affects the first 10 threads and is a taper tap.

8. Prior to drilling a hole with a twist drill, a _____ is used to create an indentation and a _____ or _____ is used to produce a starter hole to prevent the twist drill from "walking off."

    A. drive punch; countersink; centerdrill   C. press; spot face; tap drill
    B. center punch; centerdrill; spot drill   D. hole punch; gun drill; countersink

    Answer B is correct. A center punch is used to enlarge the prick punch mark made during the layout. A spot drill or centerdrill should be used to create a starter hole that will channel the drill point of the twist drill and prevent it from "walking off" the hole location.

9. The feed rate of a twist drill is specified in:

    A. TIR.

    B. RPM.

    C. IPM.

    D. IPR.

    Answer D is correct. The feed rate of a twist drill is defined as the distance the tool advances for each revolution of the drill, which is expressed in inches per revolution (IPR).

10. A tool equipped with either wire or short soft bristles and used to remove pins from a file is called a:

    A. pin remover.

    B. wire brush.

    C. file cleaner.

    D. file card.

    Answer D is correct. A file card contains short soft or wire bristles and is slid across the teeth of a file to remove pins and debris from the file.

11. Which drill would be the **best** choice to drill prior to reaming a 3/8″ Ø hole using a chucking reamer?

    A. 3/8″

    B. U

    C. 23/64″

    D. 5/16″

    Answer C is correct. The recommended material allowance for a 3/8″ Ø reamer is 0.015″. This would translate to 0.375″ – 0.015″ = 0.360″. The closest standard drill to this size is a 23/64″ (0.3594) drill.

12. Line A-A in the following drawing is a _____ line.

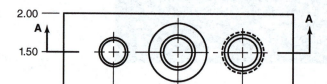

    A. leader

    B. hidden

    C. extension

    D. cutting plane

    Answer D is correct. A cutting plane line indicates an imaginary cut along this line to create a section view. A section view will provide greater clarity to internal features to be machined on the part. The cutting plane line is drawn with a long dash followed by two short dashes and has arrowheads pointing toward the viewing direction.

13. Select the semiprecision layout tool used for producing circles, radii, and arcs.

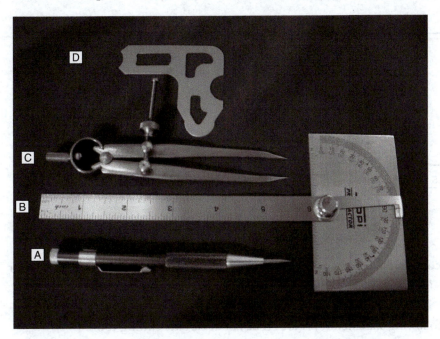

A. Scriber
B. Protractor
C. Dividers
D. Radius gage

Answer C is correct. The divider creates circles, radii, and arcs by placing one point at the center point of the geometry and pivoting the other point around that center to scribe the circle or arc.

14. The minor diameter and maximum pitch diameter for a 3/8-16 UNC-2A thread is:

A. 0.299″; 0.333″.
B. 0.321″; 334″.
C. 0.3076″; 0.3376″.
D. 0.324″; 0.347″.

Answer A is correct. Referencing a Unified Screw Thread chart in either *The Machinery's Handbook* or the *Shop Reference for Students and Apprentices*, the minor diameter for an external 3/8-16 UNC-2 thread is 0.2992″. The maximum pitch diameter for the same thread is 0.3331″.

15. Which of the following is a method of peripheral milling where the workpiece is fed with the rotation of the cutting tool?

A. Conventional milling
B. Face milling
C. Precision milling
D. Climb milling

Answer D is correct. The clockwise rotation of a cutting tool will generate a climbing action when the workpiece is fed with the rotation. This is referred to as climb milling.

16. Which angle is a solid square used to inspect?

A. 45°
B. 60°
C. 80°
D. 90°

Answer D is correct. The angle between the beam and the blade of a solid square is 90° and is therefore used to inspect perpendicularity between two surfaces.

17. What is the recommended keyseat depth for 5/8″ square key on the shaft detailed here?

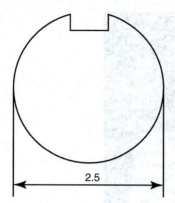

2.5

A. 0.290″   C. 0.600″

B. 0.375″   D. 0.313″

Answer D is correct. Referencing a Key Size chart in either *The Machinery's Handbook* or the *Shop Reference for Students and Apprentices*, the recommended keyseat depth is 5/16″ (0.3125).

18. Which of the following statement(s) is(are) **true** about carbide cutting tools?

A. Carbide is more expensive than high-speed steel (HSS).

B. Carbide is brittle and easily chipped.

C. Carbide tools can perform at higher speeds and feeds than HSS.

D. All of the above are true.

Answer D is correct. Tungsten carbide is very hard in comparison to HSS and offers strength and wear resistance. Tools made from carbide can perform at much higher speeds and feeds. This dense material weighs approximately twice that of HSS and is also more expensive.

19. When using a hacksaw, it is recommended to have at least _____ teeth in contact with the workpiece.

A. nine   C. five

B. seven   D. three

Answer D is correct. By having a minimum of three teeth in contact with the material being cut, the chance for a tooth catching on the edge of the metal and possibly breaking is minimized.

20. A machinist needs to cut 1/2″ off the end of a hardened steel rod (0.625 Ø × 7.5″ long). The tolerance is ± 1/16″. Select the most efficient process to complete this job.

A. Abrasive cutoff saw   C. Surface grinder

B. Wire EDM   D. Plasma torch

Answer A is correct. Though this job could be accomplished using the other processes, using an abrasive cutoff saw would be the most efficient process. With a ±1/16″ tolerance, the machinist could hold this dimension by laying out the length prior to cutting the end off.

21. When cutting harder materials with an HSS endmill, tool life can be extended and tool failure can be prevented by:

   A. increasing RPM.
   B. lowering cutting speeds.
   C. doubling feed rates.
   D. multiplying the feed rate by 1.5.

   Answer B is correct. A harder material is tougher for the tool to penetrate, resulting in increased cutting temperatures. A slower cutting speed will aid in keeping the cutting tool at a temperature that prevents tool failure.

22. _____ is commonly used in the layout process to enhance visibility of and to provide contrast to the layout lines.

   A. Spray paint
   B. Buffing
   C. Layout fluid
   D. Prussian blue

   Answer C is correct. The dark color of the layout fluid provides contrast to and enhances the visibility of the layout lines.

23. Which of the following is a result of draw filing?

   A. Smooth surface finish
   B. Fast rate of material removal
   C. Slower rate of material removal
   D. Both A and C

   Answer D is correct. Moving the file in the direction of its width will remove material at a slower rate than straight filing and will result in a smooth, finer finish.

24. Using the formula $T = L/\text{RPM} \times$ feed rate, calculate the time to machine one pass of a 0.75″ Ø × 18.04″ long, 410 stainless steel shaft at 55 SFPM, with a feed rate of 0.003 IPR. Use 4 as the constant in the RPM formula.

   A. 18 minutes
   B. 20.5 minutes
   C. 36.4 minutes
   D. 54 minutes

   Answer B is correct. Inserting the provided variables into the formula $T = L / \text{RPM} \times$ feed rate will result in $T = 18.04/293 \times 0.003$; $T = 18.04/0.879$; $T = 20.52$.

25. What is the recommended tap drill for the threaded hole on the part shown in the following figure?

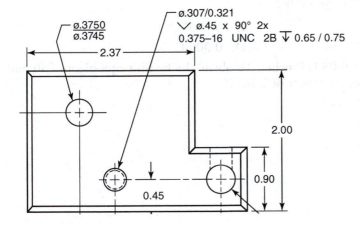

   A. 0.290″
   B. Q
   C. 3/8″
   D. 5/16″

   Answer D is correct. Referencing a Unified National Series Tap Drill chart, the specified tap drill for a 3/8-16 thread is 5/16″.

26. Emory or silicon carbide abrasive cloth is most applicable for which of the following?

    A. Polishing scratches from metal surfaces

    B. Deburring

    C. Producing a smooth surface

    D. All of the above

    E. None of the above

    Answer D is correct. Rubbing abrasive cloth over a machined part will remove burrs, while creating a smooth polished surface.

27. What is the hole diameter for a 2" Ø RC6 running fit? Use *The Machinery's Handbook* or *Shop Reference for Students and Apprentices* as a reference source.

    A. 1.9957" to 1.9975"

    B. 1.997" to 2.003"

    C. 2.000" to 2.003"

    D. 2.000" to 2.0003"

    Answer C is correct. Referencing a American National Standard Running and Sliding Fit chart in either *The Machinery's Handbook* or the *Shop Reference for Students and Apprentices*, the standard tolerance limit for this hole diameter is a range of +.000" to +.003".

28. A _____ is a layout tool used to mark lines on metal surfaces. _____ is used to enhance the visibility of layout lines.

    A. square; A shop light

    B. center punch; A scriber

    C. scriber; Layout fluid

    D. scriber; A center punch

    Answer C is correct. A scriber is a tool with a hardened steel tip and is commonly used to mark straight layout lines. The dark color of layout fluid provides contrast to and enhances the visibility of layout lines.

29. The total number of divisions on the thimble of an inch-based micrometer is:

    A. 100.

    B. 10.

    C. 15.

    D. 25.

    Answer D is correct. There are 25 lines or increments around the circumference of an inch-based micrometer, and each increment represents 0.001".

30. A machinist is given a print with metric dimensions. The conversion of a 30-mm shaft dimension to the nearest 0.0001" would be?

    A. 0.7620"

    B. 1.1811"

    C. 0.8467"

    D. 0.3000"

    Answer B is correct. 1 mm = 0.03937 inches. To obtain the English equivalent of 30 mm, multiply 30 mm × 0.03937 for an answer of 1.1811".

31. Refer to the following figure to determine how far past the 1.0″ depth it is necessary to drill, to compensate for the drill point (118° included angle drill; 0.375Ø).

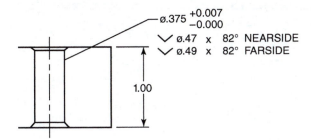

A. 0.1125″
B. 0.1875″
C. 0.2625″
D. 0.375″

Answer A is correct. To calculate the tip length of the 0.375Ø drill, use the formula tip length = 0.3 × drill Ø; tip length = 0.3 × 0.375; tip length = 0.1125″.

32. The major alloy element of 2520 alloy steel is:

A. chromium.
B. tungsten.
C. manganese.
D. nickel.

Answer D is correct. The SAE numbering system identifies the alloy element by the first two numbers of the designation. The number 25 corresponds to the alloy nickel.

33. Applying _____ to a file will aid in preventing material from becoming lodged in the teeth of the file.

A. cutting oil
B. chalk
C. WD-40
D. coolant

Answer B is correct. Applying chalk to the file's teeth will help in preventing pinnings from sticking to and clogging the file.

34. The stock allowance for a hand reaming operation is approximately:

A. 0.010″ to 0.020″.
B. 0.008″ to 0.015″.
C. 0.005″ to 0.010″.
D. 0.001″ to 0.008″.

Answer D is correct. The amount of material remaining in the hole for the hand reamer to cut should be minimal. A material allowance between 0.001″ and 0.008″ will be sufficient for providing material for the reamer to cut, while not creating too much resistance so that it would be difficult to turn it by hand.

35. Line A in the following print is a _____ line that indicates the specified geometry is _____ in the provided view.

A. hidden; not visible

B. object; not visible

C. hidden; a cross section

D. phantom; visible

Answer A is correct. A hidden line is a thin line made up of short dashes. Hidden lines show part geometry that is not visible in a particular view of the drawing.

36. Which of the following statements **best** describes a center punch?

A. 60° point angle marks the intersections of arc and circle center points.

B. 60° point angle is used to enlarge a prick punch mark for drilling.

C. 90° point angle marks the intersections of arc and circle center points.

D. 90° point angle is used to enlarge a prick punch mark for drilling.

Answer D is correct. The primary use of a center punch is to enlarge a prick punch mark to facilitate drilling. The included point angle of a center punch is 90°, whereas the point angle of a prick punch is 60°.

37. The machinist is assembling components that specify an FN2 medium drive fit. Which tool or tools should be used in this assembly process?

A. Impact wrench

B. Locking compound

C. Arbor press

D. Snap ring pliers

Answer C is correct. An FN2 medium drive fit will have a reasonable amount of interference between the mating parts. Parts with this specification can be assembled cold, and the 1/4–3 tons of force supplied by most arbor presses is sufficient for this type of assembly.

38. The vertical lines on the sleeve of a micrometer represent _____, and the horizontal lines around the circumference of a thimble are equal to _____.

A. 0.025″; 0.001″

B. 0.025″; 0.010″

C. 0.010″; 0.010″

D. 0.25″; 0.001″

Answer A is correct. An inch-based micrometer sleeve has vertical lines, each representing a 0.025″ increment. There are 25 horizontal lines around the thimble, each representing an increment of 0.001″.

39. What is the inch measurement on the vernier calipers pictured here?

A. 4.135″

B. 5.360″

C. 4.635″

D. 4.110″

Answer A is correct. To obtain the reading on a vernier scale, first determine what whole inch graduation the zero mark (on the lower scale) is past. In this case it is 4. Then determine which of the 0.100″ graduations the zero mark is past. In this case it is the 1 or 0.100″. Then determine which of the 0.025″ graduations the zero mark is past. In this case it is the first 0.025″ line. Lastly, determine which line on the lower scale is best aligned to the line above it. In this case it is the 10 or 0.010″. Add these values together to obtain the reading 4.0 + 0.100 + 0.025 + 0.010 = 4.135″.

40. What is the measurement shown on the rule pictured here?

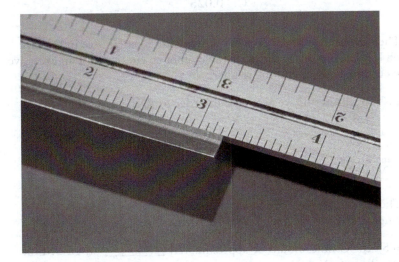

A. 3 and 3/8″

B. 3 and 3/4″

C. 3 and 3/32″

D. 3 and 3/16″

Answer D is correct. The part is being measured using the 1/16 scale of the rule. This is evident by counting the graduations between the inch increments. The edge of the part is past the 3-inch graduation mark and aligns with the third increment of the 1/16 graduations. Therefore, the part measures 3 and 3/16″.

41. A vernier caliper can accurately measure to an increment as small as:

A. 0.0010".

C. 0.0005".

B. 0.0001".

D. 0.0050".

Answer A is correct. The main scale on a vernier caliper is graduated in increments of 0.001" and can be used to accurately measure to this increment.

42. Which of the following tools should be used to finish machine hole A in the following drawing?

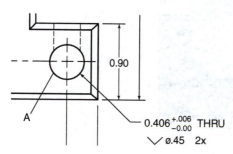

A. Reamer

C. Tap

B. Twist drill

D. Counterbore

Answer B is correct. As with any machining operation, the methods and tools used to produce part geometry are dependent on the tolerance of the particular part feature. This through (THRU) hole has a tolerance of +.006/–0.000. A twist drill that is sharpened properly and run at the correct feeds and speeds is capable of producing this hole within the specified tolerance range.

43. What is the recommended stock allowance for machine reaming a hole between 1/4" Ø and 1/2" Ø?

A. 0.015"

C. 0.025"

B. 0.0105"

D. 0.020"

Answer A is correct. The recommended material allowance for a reamed hole in this size range is 0.015". This will leave sufficient material for the reamer to cut, but not an excessive amount to overload the cutting capability of the reamer.

44. Which of the surface finish specifications are arranged in order from the smoothest to the roughest?

A. $\sqrt[4]{}$, 60 μin, $\sqrt[100]{}$, $\sqrt[150]{}$

C. 225 μin, 125 μin, 50 μin, $\sqrt[25]{}$

B. 200 μin, 150 μin, $\sqrt[100]{}$, $\sqrt[150]{}$

D. 4 μin, $\sqrt[40]{}$, $\sqrt[400]{}$, $\sqrt[150]{}$

Answer A is correct. The smaller the value in microinches, the smother the surface finish will be. Answer A is arranged sequentially from 4 to 150 microinches.

45. The RPM for reaming a 5/16" Ø hole should be approximately _____ the RPM for drilling a hole of the same diameter?

A. one-half

C. equal

B. double

D. one-fourth

Answer A is correct. If referencing a cutting speed chart, the SFPM for reaming operations is typically 50–60% that of drilling the same-size hole. If a cutting speed chart is not readily available, running the reamer at one-half the drilling speed is a reliable approximation.

46. Calculate the feed rate using the following variables: 12 flute 2.5″ Ø face mill, 65 SFPM, and 0.006″ feed per tooth. Use 4 as the constant.

    A. 7.5 IPM
    B. 4.7 IPM
    C. 46.8 IPM
    D. 8.0 IPM

    Answer A is correct. The RPM value is a variable in the feed rate calculation formula and therefore must be calculated first. Inserting the provided variables into the formula will result in RPM = 4 × 65/2.5; RPM = 260/2.5; RPM = 104. Using the formula IPM = FPT × N × RPM; IPM = 0.006 × 12 × 104; IPM = 7.488.

47. What is the minimum-diameter round stock needed to produce the wrench head specified on the following part?

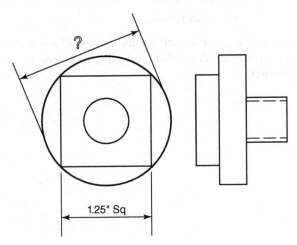

    1.25″ Sq

    A. 1.266″
    B. 1.7678″
    C. 1.5″
    D. 1.250″

    Answer B is correct. To calculate the round stock size, bisect the square by drawing a diagonal line from one corner to another. This creates a right triangle with the bisecting line as the hypotenuse. Knowing the other two legs of the triangle measure 1.25″ in length, the Pythagorean theorem can be used to solve for the hypotenuse. The Pythagorean theorem is $A^2 + B^2 = C^2$; $1.25^2 + 1.25^2 = C^2$; $3.125 = C^2$; $\sqrt{3.125} = \sqrt{C^2}$; $C = 1.7678$.

48. Dimension A specifies a _____, and dimension B specifies a _____.

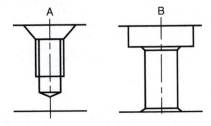

    A   B

    A. chamfer; spotface
    B. countersink; stepbore
    C. countersink; counterbore
    D. counterbore; countersink

    Answer C is correct. A is pointing to a conical or tapered opening, concentric to a through hole, to allow for the head of a flathead screw to sit flush or below the surface of the part. B is pointing to a cylindrical-shaped hole with a flat bottom and 90° corners concentric to a through hole to allow for a socket head cap screw to sit flush or below the surface of the part.

49. The distance a point on the circumference of a rotating cutting tool travels in 1 minute is expressed as:

A. revolutions per minute (RPM).

C. surface feet per minute (SFPM).

B. total indicator runout (TIR).

D. All of the above.

Answer C is correct. Cutting speed is expressed in SFPM and is the distance a point on the circumference of a rotating cutting tool travels in 1 minute. The formula RPM = 4 × CS/ diameter is used to convert SFPM to RPM.

50. A flat spot machined on a rough or angled surface to provide a bearing surface for bolts, nuts, or washers describes a:

A. pilot.

C. counterbore.

B. spotface.

D. countersink.

Answer B is correct. A rough surface surrounding a hole used for fastening will not provide ample surface contact against the nut, bolt, or washer. A spotface is machined on the part to create a smooth bearing surface for the fastener.

Drill Press Certification

4

4.1   PRACTICE THEORY EXAM—DRILL PRESS

1. A machinist reamed a 0.375 Ø hole in 1030 plain carbon steel. The no-go pin passes through the hole. Which of the following may have caused this to occur?

    A. The RPM were 60% slower than the drilling cutting speed.

    B. The reamer's cutting edges were worn.

    C. The reamer was new and was not "broken in."

    D. Excessive material was left for reaming.

2. A countersink is often used to machine a _____ on the opening of a hole to provide "lead-in" for taps, reamers, and dowel pins.

    A. filet

    B. chamfer

    C. cone

    D. recess

3. Calculating and setting the correct RPM for a drill press operation is essential for ensuring safety because:

    A. a slower running drill produces smaller chips.

    B. a slower RPM means it is OK to hold the part by hand.

    C. using the optimal RPM will result in minimal tool wear and lessen the chance of drill breakage.

    D. all of the above ensure safety.

4. Which tool pictured here would be used to enlarge a hole diameter to a specific depth to allow a flathead screw to sit flush or below the part's surface?

    A. A

    B. B

    C. C

    D. D

5. Which of the following combinations of variables will have the most effect on a drilled hole diameter?

   A. The lubrication, the RPM, the accuracy the point is ground to, the feed rate

   B. The workholding device, the length of the drill, the point angle, the lubricant

   C. The material being cut, the drill material, the feed rate, the length of the drill

   D. The horsepower of the drill press, the lubrication, the material being cut, the length of the drill

6. A part print specifies two tapped holes. The first is a 5/16-24 UNC-2B, and the second is a 3/8-24 UNC-2B. What does 24 represent?

   A. Class of fit

   B. Minor diameter

   C. Form angle

   D. Threads per inch

7. A quality control inspector rejects several parts because a drilled hole is oversize. Which answer **best** identifies the possible cause of this problem?

   A. The point angle is ground incorrect.

   B. The lips are ground to different lengths.

   C. The drill is made of carbide.

   D. Both A and B are correct.

   E. None of the above is correct.

8. Rapid tool wear, blue chips, and excessive tool chatter when drilling on a drill press with high-speed steel are indicators of:

   A. deep drilling.

   B. slow feed rate.

   C. soft material.

   D. too fast spindle speed.

9. Used to create a "start" for the tip of a twist drill, a _____ is a short drill and countersink combination with a 60° included angle on the countersink portion, or a _____, which is a short drill with a point angle ranging from 90° to 120°, should be used.

   A. countersink; spot face

   B. centerdrill; spot drill

   C. spot drill; centerdrill

   D. spot face; countersink

10. Which is the machining operation usually performed most frequently on a drill press and is often a requirement prior to other drill press machining operations?

    A. Boring

    B. Countersinking

    C. Tapping

    D. Drilling

11. The finished diameter of a drilled hole is typically:

    A. 0.003"–0.005" undersized.

    B. 0.005"–0.010" oversized.

    C. on size.

    D. 0.003"–0.005" oversized.

12. Letter **J** in the following picture identifies which part of the drill press?

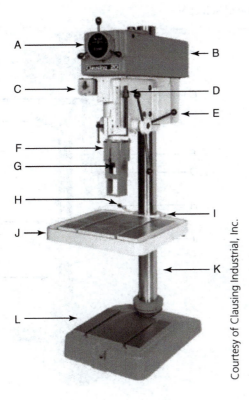

Courtesy of Clausing Industrial, Inc.

A. Table   C. Column
B. Base   D. Work surface

13. Letter **A** in the following picture identifies which part of the drill press?

    A. Feed handle   C. Variable speed selector

    B. Table elevating crank   D. Quill feed

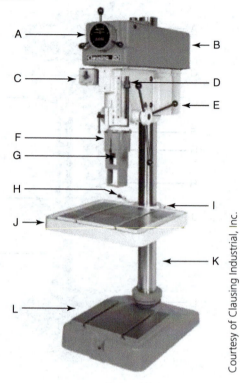

Courtesy of Clausing Industrial, Inc.

14. Letter **D** in the following picture identifies which part of the drill press?

    A. Quill   C. Depth stop

    B. Spindle   D. Table clamp

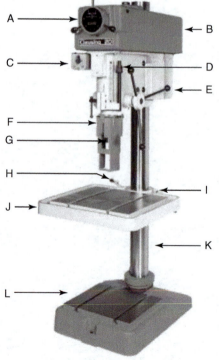

Courtesy of Clausing Industrial, Inc.

15. The **spindle** in the following picture is identified by which letter?

    A. E   C. C
    B. B   D. G

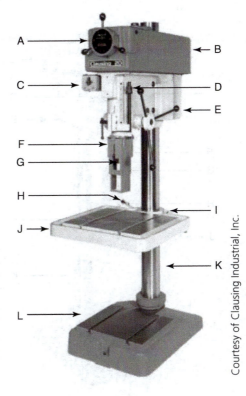

Courtesy of Clausing Industrial, Inc.

16. The **column** in the following picture is identified by which letter?

    A. K   C. F
    B. G   D. E

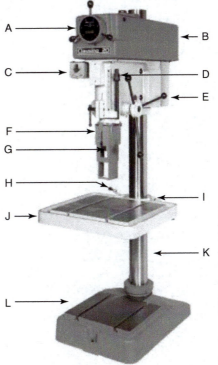

Courtesy of Clausing Industrial, Inc.

17. The **quill feed handle** in the following picture is identified by which letter?

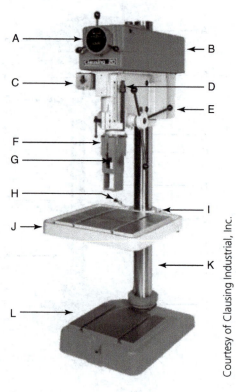

Courtesy of Clausing Industrial, Inc.

A. I   C. H
B. E   D. A

18. The **drive belts and pulleys** are covered by the component identified by which letter in the following picture?

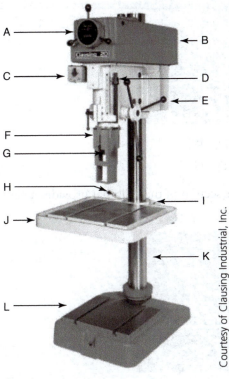

Courtesy of Clausing Industrial, Inc.

A. A   C. C
B. B   D. E

19. Which letter in the following picture identifies the shank of the drill?

A. D   C. F
B. E   D. G

20. Which letter in the following picture identifies the margin of the drill?

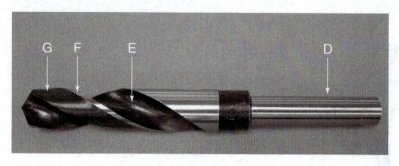

A. D

B. E

C. F

D. G

21. Which letter in the following picture identifies the section of the drill that provides a pathway for the chips to evacuate during drilling operations?

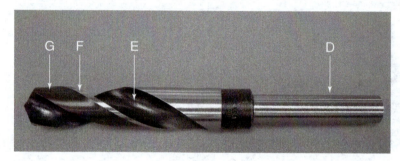

A. D

B. E

C. F

D. G

22. Which letter in the following picture identifies the land of a twist drill?

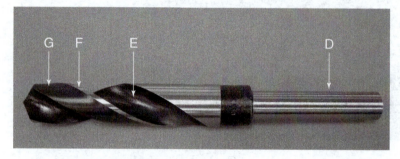

A. D

B. E

C. F

D. G

23. Which of the following techniques and procedures are recommended when machining cast iron?

    A. Wear a dust mask to avoid inhaling cast iron dust.

    B. Take precautionary steps to protect the machine from the cast iron dust and chips.

    C. Cutting oil is typically not necessary.

    D. All of the above are recommended.

    E. None of the above is recommended.

24. The metal chips produced while operating a drill press should be handled as a:

    A. safety hazard.

    B. by-product.

    C. waste of material.

    D. consequence of machining.

25. A machinist has just finished using a drill press to produce a part and does the following. Which of these actions is considered safe and acceptable?

    A. Uses his hand to stop the spindle rotation

    B. Wipes the chips off of the vise with his hand

    C. Removes the remaining chips with a brush

    D. Leaves the chuck key in the socket

    E. Decides to leave a cutting oil spill for cleanup at the end of the day

26. To create the flat bearing surface for the head of the socket cap screw in the following image, a _____ must be machined on the workpiece.

    A. pilot

    B. shoulder

    C. counterbore

    D. spotface

27. Referencing *The Machinery's Handbook* or the *Shop Reference for Students and Apprentices*, find which tap drill should be used for a 3/4-14 NPT.

    A. 23/32″

    B. 59/64″

    C. 11/16″

    D. 21/32″

28. The print specifies a 0.375 Ø hole to be reamed through. Which of the following drills should be used to drill the hole prior to reaming?

    A. T (0.3580)

    B. 3/8 (0.375)

    C. U (0.368)

    D. 23/64 (0.3594)

29. While reaming a series of holes on the drill press, the machinist has cutting oil splashed in her eye. Where would she find first-aid information about this situation?

    A. On a material safety data sheet

    B. On the back of the container label

    C. Job router sheet

    D. OSHA Web site

30. A drill point gage is an inspection tool used to:

    A. check the helix angle of a drill.

    B. verify the lip clearance.

    C. measure the drill point angle and lip length.

    D. measure the diameter of the drill.

31. Which of the following is correct about IPR?

    A. It specifies the feed rate on a drill press.

    B. It is the acronym for inches per revolution.

    C. It specifies the distance a cutting tool will advance per one complete revolution of the spindle.

    D. All of the above are correct.

    E. None of the above is correct.

32. What is likely to occur when drilling a 49/64″ Ø hole in the piece pictured in the following setup?

    A. The drill will cut straight through the part efficiently.

    B. The part will flex due to feed pressure and could spring back during "breakthrough," causing the drill to break.

    C. The part will flex due to feed pressure, and the hole may not be perpendicular to the top of the workpiece.

    D. B and C are correct.

    E. None of the above is correct.

33. The print specifies a reamed hole dimension at 0.500″ + 0.001″, −0.000. Which of the following methods could be used to inspect the size of this dimension?

    A. Dial calipers

    B. Telescope gage

    C. 0.500″ and 0.502″ gage pin

    D. 0.499″ and 0.501″ gage pin

34. A large drill press with a heavy base designed for drilling large-diameter holes in large workpieces is a:

    A. column drill press.

    B. upright drill press.

    C. vertical drill press.

    D. radial drill press.

35. A standard twist drill will have an included point angle of:

    A. 135°.

    B. 100°.

    C. 128°.

    D. 118°.

36. Which of the following could lead to an injury when operating a drill press?

    A. Wearing loose, baggy clothing
    B. Long unsecured hair
    C. Wearing a watch
    D. Wearing your ID on a lanyard around your neck
    E. All of the above

37. A prick punch has a _____ point angle and is used to mark the center points of circles, arcs, and radii. A center punch has a _____ point angle and is used to enlarge a prick punch mark prior to drilling.

    A. 60°; 90°
    B. 45°; 90°
    C. 90°; 60°
    D. 45°; 118°

38. To drill a cross hole through a round workpiece, which of the following workholding methods can be used?

    A. V-block clamped to the table
    B. A vise with "V" slots ground in the jaws
    C. A vise with the centerline of the workpiece positioned below the top of the vise jaws
    D. All of the above
    E. A and C

39. The cutting speed of a drilling operation is calculated at 760 RPM; the approximate cutting speed for reaming a hole of this size is:

    A. 240.
    B. 360.
    C. 480.
    D. 500.

40. Which of the following will **not** aid in preventing centerdrill breakage?

    A. Apply heavy feed.
    B. Apply cutting fluid.
    C. Do not feed the body into the workpiece.
    D. Frequently retract from the work.

41. The feed rate of a drilling operation is calculated at 0.006 IPR; the approximate feed rate for reaming this hole is:

    A. 0.003.
    B. 0.006.
    C. 0.009.
    D. 0.012.

42. It is often necessary to grind a drill point nearly flat when drilling materials like brass or bronze. The nearly flat drill point will:

    A. eliminate the need for centerdrill.
    B. reduce drill runout.
    C. reduce the capability of the drill to grab at breakthrough.
    D. allow for faster RPM.

43. The blue-shaded portion of the following image is the _____ of the drill, which becomes thicker as a twist drill is ground shorter.

   A. web   C. margin

   B. dead center   D. core

44. The material required for the following part is 1060 plain carbon steel with a Brinell hardness of 200. What RPM should be used for reaming hole C in the following drawing? (Use 3.82 as the constant.)

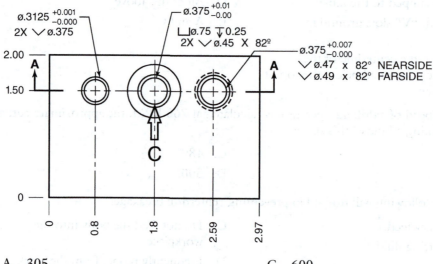

   A. 305   C. 600

   B. 458   D. 550

45. A flared or oval shape at the beginning of a reamed hole describes a condition called _____ and is caused by_____.

   A. taper; excessive feed rate   C. lead-in; dull cutting edges

   B. bellmouth; a misaligned reamer or tool holder   D. total runout; too little material left for reaming

46. What is the approximate percentage of thread achieved when using tap drill sizes recommended by most Unified National Tap Drill charts?

   A. 100%   D. 65%

   B. 85%   E. None of the above

   C. 75%

47. The depth of an internal thread (thread percentage) tapped on a drill press is controlled by the:

   A. material.   C. tap drill diameter.

   B. cutting oil.   D. length of the tap drill.

48. A print calls for ten 0.033"-diameter holes to be drilled through 1/2"-thick O2 steel. Which of the following answers could help to prevent the drill from breaking?

   A. Apply double quantities of cutting oil.   C. Decrease the feed rate.

   B. Decrease the RPM.   D. Increase the RPM.

49. Which of the following reamers is recommended for interrupted cuts such as reaming past a cross hole?

   A. Expansion reamer   C. Straight machine reamer

   B. Spiral machine reamer   D. Taper pin reamer

50. The following drawing has five holes equally spaced. What is the distance between each hole?

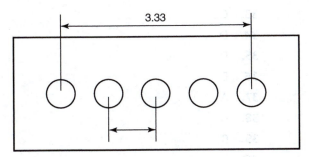

   A. 0.6665"   C. 0.6255"

   B. 0.3330"   D. 0.8325"

4.2   ANSWER KEY

1.	D	26.	D
2.	B	27.	B
3.	C	28.	D
4.	B	29.	A
5.	A	30.	C
6.	D	31.	D
7.	D	32.	D
8.	D	33.	C
9.	B	34.	D
10.	D	35.	D
11.	D	36.	E
12.	A	37.	A
13.	C	38.	D
14.	C	39.	B
15.	D	40.	A
16.	A	41.	D
17.	B	42.	C
18.	B	43.	A
19.	A	44.	B
20.	C	45.	B
21.	B	46.	C
22.	D	47.	C
23.	D	48.	C
24.	A	49.	B
25.	C	50.	D

4.3 PRACTICE THEORY EXAM EXPLANATIONS

1. A machinist reamed a 0.375 Ø hole in 1030 plain carbon steel. The no-go pin passes through the hole. Which of the following may have caused this to occur?

   A. The RPM were 60% slower than the drilling cutting speed.

   B. The reamer's cutting edges were worn.

   C. The reamer was new and was not "broken in."

   D. Excessive material was left for reaming.

   Answer D is correct. Excessive material may cause the flutes of the reamer to become loaded with chips, resulting in the reamer pushing off the sides of the hole. This pushing effect could cause the reamer to cut more material from the opposing side, resulting in an oversize hole.

2. A countersink is often used to machine a _____ on the opening of a hole to provide "lead-in" for taps, reamers, and dowel pins.

   A. filet

   B. chamfer

   C. cone

   D. recess

   Answer B is correct. One of the uses of a countersink tool is to machine chamfers on existing holes.

3. Calculating and setting the correct RPM for a drill press operation is essential for ensuring safety because:

   A. a slower running drill produces smaller chips.

   B. a slower RPM means it is OK to hold the part by hand.

   C. using the optimal RPM will result in minimal tool wear and lessen the chance of drill breakage.

   D. all of the above ensure safety.

   Answer C is correct. A spindle speed that is too fast will lead to excessive heat or wear to the cutting tool. This accelerated wear could result in damage to the tool and workpiece during the machining operation.

4. Which tool pictured here would be used to enlarge a hole diameter to a specific depth to allow a flathead screw to sit flush or below the part's surface?

   A. A

   B. B

   C. C

   D. D

   Answer B is correct. One of the uses of a countersink tool is to machine a countersink in the workpiece to provide a recess for a flathead screw.

5. Which of the following combinations of variables will have the most effect on a drilled hole diameter?

   A. The lubrication, the RPM, the accuracy the point is ground to, the feed rate

   B. The workholding device, the length of the drill, the point angle, the lubricant

   C. The material being cut, the drill material, the feed rate, the length of the drill

   D. The horsepower of the drill press, the lubrication, the material being cut, the length of the drill

   Answer A is correct. This combination will have the greatest effect on the efficiency and accuracy of the cutting action and the finished diameter. Lubrication is essential for reducing the impact of friction and heat, while providing longer tool life. Correct RPM and feed rate will provide optimal cutting efficiency and also reduce tool wear. Proper tool geometry will ensure the drill cuts "on-center."

6. A part print specifies two tapped holes. The first is a 5/16-24 UNC-2B, and the second is a 3/8"-24 UNC-2B. What does 24 represent?

   A. Class of fit

   B. Minor diameter

   C. Form angle

   D. Threads per inch

   Answer D is correct. The Unified Screw Thread Standard identifies the threads per inch with this specification in a thread designation.

7. A quality control inspector rejects several parts because a drilled hole is oversize. Which answer **best** identifies the possible cause of this problem?

   A. The point angle is ground incorrect.

   B. The lips are ground to different lengths.

   C. The drill is made of carbide.

   D. Both A and B are correct.

   E. None of the above is correct.

   Answer D is correct. If the separate cutting lips are ground at different angles to the axis of rotation or the lengths of the cutting lips are unequal, resulting in the dead center being misaligned to the axis of rotation, the drill will cut oversize.

8. Rapid tool wear, blue chips, and excessive tool chatter when drilling on a drill press with high-speed steel are indicators of:

   A. deep drilling.

   B. slow feed rate.

   C. soft material.

   D. too fast of a spindle speed.

   Answer D is correct. Excessive RPM will result in greater thermal buildup in the workpiece and tool, which could result in discolored chips, rapid tool wear, and eventually chatter, from dull cutting conditions.

9. Used to create a "start" for the tip of a twist drill, a _____ is a short drill and countersink combination with a 60° included angle on the countersink portion, or a _____, which is a short drill with a point angle ranging from 90° to 120°, should be used.

   A. countersink; spot face

   B. centerdrill; spot drill

   C. spot drill; centerdrill

   D. spot face; countersink

   Answer B is correct. A centerdrill is a drill and countersink combination with a 60° included angle on the countersink portion. A spot drill is a short drill bit with a 90°–120° tip angle. Both are used to create a channel for the dead center of a twist drill to prevent it from "walking" as it begins to penetrate the workpiece.

10. Which is the machining operation usually performed most frequently on a drill press and is often a requirement prior to other drill press machining operations?

    A. Boring
    B. Countersinking
    C. Tapping
    D. Drilling

    Answer D is correct. The drill press is capable of performing several holemaking operations; however, drilling is performed most frequently on this machine tool.

11. The finished diameter of a drilled hole is typically:

    A. 0.003″–0.005″ undersized.
    B. 0.005″–0.010″ oversized.
    C. on size.
    D. 0.003″–0.005″ oversized.

    Answer D is correct. Factors such as point angle, material, lubrication, the condition of the cutting edges, and drill size will dictate how much oversize a drilled hole will be, but a twist drill will typically produce a hole slightly larger than its nominal diameter.

12. Letter **J** in the following picture identifies which part of the drill press?

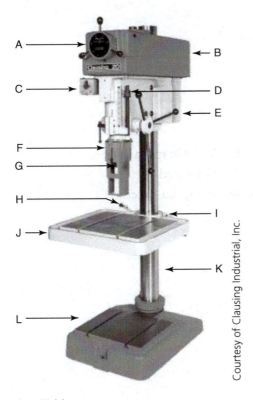

Courtesy of Clausing Industrial, Inc.

    A. Table
    B. Base
    C. Column
    D. Work surface

    Answer A is correct. Letter J is pointing to the drill press table.

13. Letter **A** in the following picture identifies which part of the drill press?

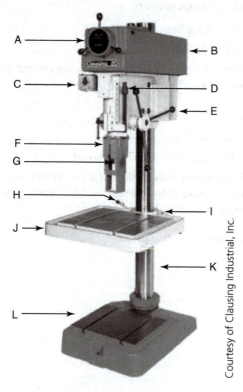

Courtesy of Clausing Industrial, Inc.

A. Feed handle   C. Variable speed selector
B. Table elevating crank   D. Quill feed

Answer C is correct. Letter A is pointing to the variable speed selector.

14. Letter **D** in the following picture identifies which part of the drill press?

   A. Quill
   B. Spindle
   C. Depth stop
   D. Table clamp

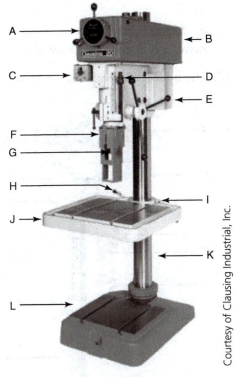

Courtesy of Clausing Industrial, Inc.

Answer C is correct. Letter D is pointing to the adjustable quill depth stop.

15. The **spindle** in the following picture is identified by which letter?

   A. E
   B. B
   C. C
   D. G

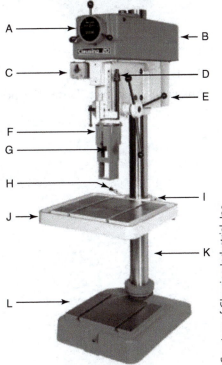

Courtesy of Clausing Industrial, Inc.

Answer D is correct. Letter G is pointing to the spindle of the drill press.

16. The **column** in the following picture is identified by which letter?

    A. K

    C. F

    B. G

    D. E

    Answer A is correct. Letter K is pointing to the column of the drill press.

17. The **quill feed handle** in the following picture is identified by which letter?

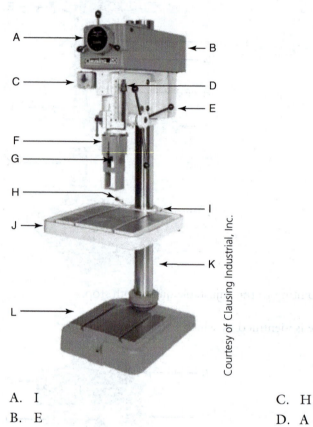

Courtesy of Clausing Industrial, Inc.

    A. I

    C. H

    B. E

    D. A

    Answer B is correct. Letter E is pointing to the quill feed handle.

18. The **drive belts and pulleys** are covered by the component identified by which letter in the following picture?

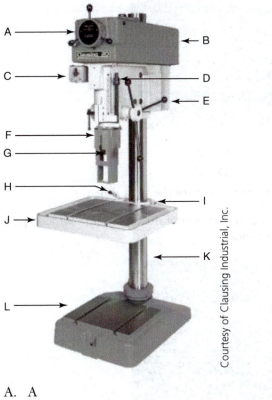

Courtesy of Clausing Industrial, Inc.

A. A
B. B

C. C
D. E

Answer B is correct. Letter B is pointing to the belt and pulley guard.

19. Which letter in the following picture identifies the shank of the drill?

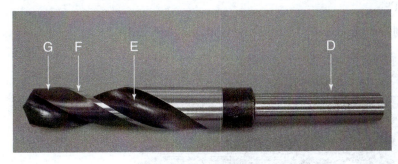

A. D
B. E

C. F
D. G

Answer A is correct. Letter D is pointing to the shank of the drill.

20. Which letter in the following picture identifies the margin of the drill?

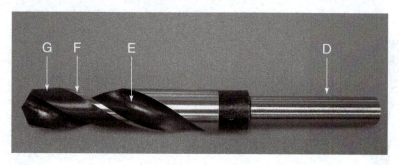

A. D

B. E

C. F

D. G

Answer C is correct. Letter F is pointing to the margin of the twist drill.

21. Which letter in the following picture identifies the section of the drill that provides a pathway for the chips to evacuate during drilling operations?

A. D

B. E

C. F

D. G

Answer B is correct. Letter E is pointing to one of the flutes on the drill, which provides a path for chips to escape.

22. Which letter in the following picture identifies the land of a twist drill?

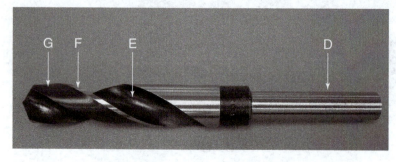

A. D

B. E

C. F

D. G

Answer D is correct. Letter G is pointing to the land of the twist drill.

23. Which of the following techniques and procedures are recommended when machining cast iron?

    A. Wear a dust mask to avoid inhaling cast iron dust.
    B. Take precautionary steps to protect the machine from the cast iron dust and chips.
    C. Cutting oil is typically not necessary.
    D. All of the above are recommended.
    E. None of the above is recommended.

    Answer D is correct. The rough outer scale and machinability of cast iron necessitates precautionary measures to protect the machinist and the equipment. The internal properties and heat conductivity of cast iron all but eliminate any advantage of using cutting oil.

24. The metal chips produced while operating a drill press should be handled as a:

    A. safety hazard.
    B. by-product.
    C. waste of material.
    D. consequence of machining.

    Answer A is correct. Metal chips have the potential to cause injury and should be handled as potentially hazardous.

25. A machinist has just finished using a drill press to produce a part and does the following. Which of these actions is considered safe and acceptable?

    A. Uses his hand to stop the spindle rotation
    B. Wipes the chips off of the vise with his hand
    C. Removes the remaining chips with a brush
    D. Leaves the chuck key in the socket
    E. Decides to leave a cutting oil spill for cleanup at the end of the day

    Answer C is correct. To safely remove metal chips from a machine tool, a brush or rag should be used.

26. To create the flat bearing surface for the head of the socket cap screw in the following image, a _____ must be machined on the workpiece.

    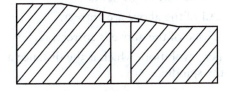

    A. pilot
    B. shoulder
    C. counterbore
    D. spotface

    Answer D is correct. A spotface creates a flat bearing surface for a fastener to seat when tightened.

27. Referencing *The Machinery's Handbook* or the *Shop Reference for Students and Apprentices*, find which tap drill should be used for a 3/4-14 NPT.

    A. 23/32"   C. 11/16"
    B. 59/64"   D. 21/32"

    Answer B is correct. Referencing a Unified National Series Tap Drill chart, the tap drill specified for a 3/4-14 NPT is 59/64".

28. The print specifies a 0.375 Ø hole to be reamed through. Which of the following drills should be used to drill the hole prior to reaming?

    A. T (0.3580)   C. U (0.368)
    B. 3/8 (0.375)   D. 23/64 (0.3594)

    Answer D is correct. The recommended stock allowance for a 0.375 Ø is 0.015". A 23/64" drill is the closest standard twist drill to this size.

29. While reaming a series of holes on the drill press, the machinist has cutting oil splashed in her eye. Where would she find first-aid information about this situation?

    A. On a material safety data sheet   C. Job router sheet
    B. On the back of the container label   D. OSHA Web site

    Answer A is correct. A material safety data sheet will contain a section on first-aid measures.

30. A drill point gage is an inspection tool used to:

    A. check the helix angle of a drill.   C. measure the drill point angle and lip length.
    B. verify the lip clearance.   D. measure the diameter of the drill.

    Answer C is correct. A drill point gage is used to verify that the correct angle is ground on the point of a drill and the cutting lips have been ground to the same length.

31. Which of the following is correct about IPR?

    A. It specifies the feed rate on a drill press.   tool will advance per one complete revolution of the spindle.
    B. It is the acronym for inches per revolution.   D. All of the above are correct.
    C. It specifies the distance a cutting   E. None of the above is correct.

    Answer D is correct. IPR is the acronym for inches per revolution, which is the distance a drill bit will advance per one complete revolution of the spindle.

32. What is likely to occur when drilling a 49/64″ Ø hole in the piece pictured in the following setup?

A. The drill will cut straight through the part efficiently.

B. The part will flex due to feed pressure and could spring back during "breakthrough," causing the drill to break.

C. The part will flex due to feed pressure, and the hole may not be perpendicular to the top of the workpiece.

D. B and C are correct.

E. None of the above is correct.

Answer D is correct. The part is not supported near the drilled hole location, which will allow the part to flex during drilling. At the point the drill breaks through, the piece may spring back, resulting in the thin remaining material in the hole to "pop" out the bottom. The ending result could be a large burr or "fin" on the bottom of the hole and possible binding of the drill upon the spring back. In addition, the material movement will likely produce a hole that is not perpendicular to the top surface of the part.

33. The print specifies a reamed hole dimension at 0.500″ + 0.001″, –0.000. Which of the following methods could be used to inspect the size of this dimension?

A. Dial calipers

B. Telescope gage

C. 0.500″ and 0.502″ gage pin

D. 0.499″ and 0.501″ gage pin

Answer C is correct. The upper limit of this feature is 0.501″ and the lower limit is 0.500″. The 0.500″ gage pin would be the go pin, and the 0.502″ pin would be the no-go pin.

34. A large drill press with a heavy base designed for drilling large-diameter holes in large workpieces is a:

A. column drill press.

B. upright drill press.

C. vertical drill press.

D. radial drill press.

Answer D is correct. A radial-arm drill press has a large arm that is fitted around a large column. The spindle is mounted on the arm and can slide along the length of the arm. The arm can be raised and lowered as well as rotated 360° around the column.

35. A standard twist drill will have an included point angle of:

A. 135°.

B. 100°.

C. 128°.

D. 118°.

Answer D is correct. For most materials, 118° is the recommended point angle for general drilling operations.

36. Which of the following could lead to an injury when operating a drill press?

    A. Wearing loose, baggy clothing
    B. Long unsecured hair
    C. Wearing a watch
    D. Wearing your ID on a lanyard around your neck
    E. All of the above

    Answer E is correct. When operating machine tools with rotating tools and other moving components, long hair, clothing, and jewelry can get caught and pull the operator into the machine or cutting tool, resulting in serious injury.

37. A prick punch has a _____ point angle and is used to mark the center points of circles, arcs, and radii. A center punch has a _____ point angle and is used to enlarge a prick punch mark prior to drilling.

    A. 60°; 90°
    B. 45°; 90°
    C. 90°; 60°
    D. 45°; 118°

    Answer A is correct. A prick punch has an included tip angle of 60°, and a center punch has an included tip angle of 90°.

38. To drill a cross hole through a round workpiece, which of the following workholding methods can be used?

    A. V-block clamped to the table
    B. A vise with "V" slots ground in the jaws
    C. A vise with the centerline of the workpiece positioned below the top of the vise jaws
    D. All of the above
    E. A and C

    Answer D is correct. Each of these workholding devices will provide three points of contact on the workpiece and are applicable for this operation.

39. The cutting speed of a drilling operation is calculated at 760 RPM; the approximate cutting speed for reaming a hole of this size is:

    A. 240.
    B. 360.
    C. 480.
    D. 500.

    Answer B is correct. As a "rule of thumb," the cutting speed for a reaming operation should be approximately half of the same-size drilling operation. The closest RPM option is 360 RPM.

40. Which of the following will **not** aid in preventing centerdrill breakage?

    A. Apply heavy feed.
    B. Apply cutting fluid.
    C. Do not feed the body into the workpiece.
    D. Frequently retract from the work.

    Answer A is correct. A centerdrill has a relatively small and delicate tip. Excessive feed pressure will increase the propensity of its breaking.

41. The feed rate of a drilling operation is calculated at 0.006 IPR; the approximate feed rate for reaming this hole is:

    A. 0.003.
    B. 0.006.
    C. 0.009.
    D. 0.012.

    Answer D is correct. As a "rule of thumb," the feed rate for a reaming operation should be approximately two times the feed rate of the same-size drilling operation.

42. It is often necessary to grind a drill point nearly flat when drilling materials like brass or bronze. The nearly flat drill point will:

A. eliminate the need for centerdrill.

B. reduce drill runout.

C. reduce the capability of the drill to grab at breakthrough.

D. allow for faster RPM.

Answer C is correct. The flat bottom of the drill will allow the entire drill diameter to break through the bottom at the same time, which minimizes the opportunity for the drill to grab.

43. The blue-shaded portion of the following image is the _____ of the drill, which becomes thicker as a twist drill is ground shorter.

A. web

B. dead center

C. margin

D. core

Answer A is correct. The web runs through the center of the drill connecting the two flutes. It is tapered and increases in thickness as it reaches the shank.

44. The material required for the following part is 1060 plain carbon steel with a Brinell hardness of 200. What RPM should be used for reaming hole C in the following drawing? (Use 3.82 as the constant.)

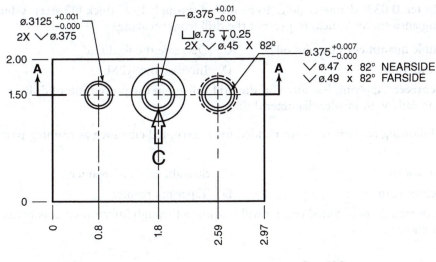

A. 305

B. 458

C. 600

D. 550

Answer B is correct. Inserting the provided variables into the RPM formula will result in

$$RPM = \frac{3.82 \times 45}{.375} = 458.4.$$

45. A flared or oval shape at the beginning of a reamed hole describes a condition called _____ and is caused by_____.

A. taper; excessive feed rate
C. lead-in; dull cutting edges
B. bellmouth; a misaligned reamer or tool holder
D. total runout; too little material left for reaming

Answer B is correct. The eccentric cutting action caused by the reamer runout will produce a larger opening at the top of the hole.

46. What is the approximate percentage of thread achieved when using tap drill sizes recommended by most Unified National Tap Drill charts?

A. 100%
D. 65%
B. 85%
E. None of the above
C. 75%

Answer C is correct. The majority of tap drill charts recommend tap drill sizes that provide a thread percentage ranging from 51% to 80%, with an average of 75%.

47. The depth of an internal thread (thread percentage) tapped on a drill press is controlled by the:

A. material.
C. tap drill diameter.
B. cutting oil.
D. length of the tap drill.

Answer C is correct. To an extent, a tap drill smaller than what is recommended by a drill chart will provide more material in the hole, increasing the thread percentage. A tap drill larger than what is recommended by a drill chart will decrease the material in the hole, therefore decreasing the thread percentage.

48. A print calls for ten 0.033"-diameter holes to be drilled through 1/2"-thick O2 steel. Which of the following answers could help to prevent the drill from breaking?

A. Apply double quantities of cutting oil.
C. Decrease the feed rate.
B. Decrease the RPM.
D. Increase the RPM.

Answer C is correct. Applying less force on the drill will aid in the prevention of drill breakage, particularly with smaller-diameter drills.

49. Which of the following reamers is recommended for interrupted cuts such as reaming past a cross hole?

A. Expansion reamer
C. Straight machine reamer
B. Spiral machine reamer
D. Taper pin reamer

Answer B is correct. A spiral fluted reamer will transition through interrupted cuts better than the other choices.

50. The following drawing has five holes equally spaced. What is the distance between each hole?

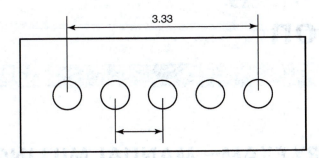

A. 0.6665″   C. 0.6255″

B. 0.3330″   D. 0.8325″

Answer D is correct. Distance between holes = 3.33″ ÷ 4 = 0.8325″.

5.1 PRACTICE THEORY EXAM— MANUAL MILLING

1. Which tool could be used to measure the width of a 0.500″ slot milled in a workpiece?

   A. Dial calipers

   B. Depth micrometer

   C. 0″–1″ micrometer

   D. Anvil micrometer

2. The chip load per tooth for a 1/2″-Ø, two-flute endmill is 0.003″. What is the calculated feed rate in inches per revolution to plunge cut with this endmill?

   A. 0.003″

   B. 0.006″

   C. 0.012″

   D. 0.024″

3. Which of the following scenarios could result in breaking the tap?

   A. Prior to tapping, inspect the tap for worn, dull, or chipped cutting edges.

   B. If the recommended tap drill is unavailable, use a smaller-size drill.

   C. Chamfer the hole prior to tapping.

   D. Apply a generous amount of cutting oil.

   E. None of the above could result in breaking the tap.

4. The micrometer collars on the X-, Y-, and Z- (knee) axes of a vertical milling machine are graduated in _____ increments.

   A. 0.010″

   B. 0.005″

   C. 0.001″

   D. 0.0005″

5. Which of the following answers is part of the process for aligning a milling vise parallel to the X-axis on a vertical milling machine?

   A. Assure the vise and table are clean and burr free.

   B. Run a dial indicator across the moveable jaw.

   C. Run a dial indicator across the solid jaw.

   D. Tap the vise into alignment with a soft-face hammer.

   E. A, C, and D are correct.

6. A machinist is required to machine the angle specified on the following part print. Which setup method would be applicable for this operation?

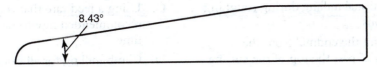

A. Hold the workpiece in a vise and tilt the head to the required angle.

B. Use a sine bar to set the angle and clamp the part to an angle block along the length of the workpiece.

C. Clamp the piece in a vise with the angled surface extending from the side of the vise and use the graduations on the swivel base to set the angle.

D. None of the above would be applicable for this operation.

7. The _____ is a component on a vertical milling machine that pulls the tool holder into the R-8 taper located on the inside of the spindle.

A. retention rod

B. spindle clamp

C. pull stud

D. drawbar

8. The process of aligning the head on a vertical milling machine perpendicular to the table in the X- and Y-axes is called:

A. tramming.

B. squaring.

C. milling.

D. quill locating.

9. To align the vertical centerline of the woodruff keyseat cutter with the vertical centerline of the shaft in the following image, what adjustment to the machine is necessary? The shaft diameter is 1.0″ and the woodruff cutter is 0.250″ thick.

A. The table should be raised 0.500″.

B. The table should be raised 0.500″ + 0.125″.

C. The table should be raised 0.375″.

D. The table should be raised 0.500″ + 0.250″.

10. The print specifies 5/16″ radii in the four corners of a manually milled pocket. What size endmill will be necessary when finishing machining this pocket to produce the specified radii?

A. 0.250″ Ø

B. 0.3125″ Ø

C. 0.4375″ Ø

D. 0.625″ Ø

11. A machinist is producing a part on a vertical milling machine with no power feed. Which of the following methods is considered safe and acceptable?

    A. Conventional milling during roughing operations

    B. Extending the endmill from the collet 1/2″ past the top of the cutting edges

    C. Using a feed rate that is twice the recommended rate to save machining time

    D. Climb milling for all side milling operations

12. Which of the following variables has the greatest effect on determining the RPM for a milling operation?

    A. The number of flutes on the cutter

    B. The horsepower of the machine

    C. The tolerance of the part features

    D. The material being machined

13. A part print specifies a hole diameter as 19/64″ (0.2969) ± 0.001″. Which combination of gage pins could be used as a go/no-go gage for this dimension?

    A. 0.295″ and 0.299″

    B. 0.296″ and 0.298″

    C. 0.295″ and 0.298″

    D. None of these

14. To machine hole "A" on the following part print and hold the location specifications, which method would be most accurate?

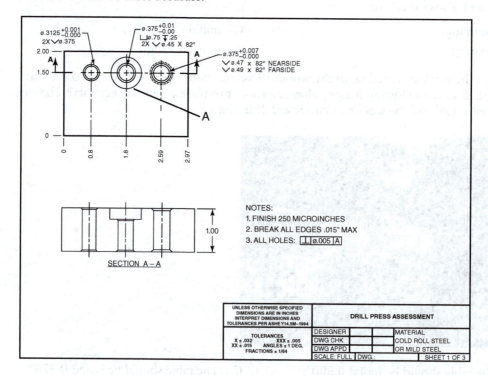

    A. Lay out the hole locations with a height gage, center punch, use a hand drill to centerdrill, and drill.

    B. Lay out the hole locations with a height gage, center punch, set up on a drill press, use a center finder to locate the position, centerdrill, and drill.

    C. Set up the workpiece on a vertical mill, locate the workpiece origin with an edge finder, use the DRO to position the table at the hole location, centerdrill, and drill.

    D. Lay out the hole location, set up the piece on a vertical milling machine, use a cone-pointed edge finder to locate the position, spot drill, and drill.

15. What type of tap should be used to produce the thread to the required depth on hole "B"?

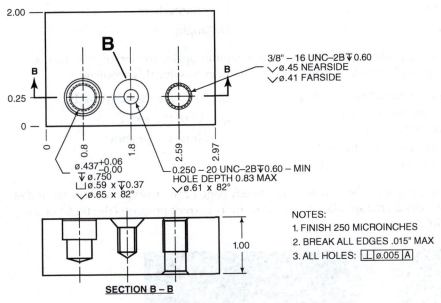

A. Taper
B. Plug
C. Bottoming
D. Spiral

16. The standard taper inside the spindle of a vertical milling machine, which is used to center the tool holder, is known as a:

A. 5C.
B. 29°.
C. Morse.
D. R-8.

17. A modern milling machine equipped with a variable speed control utilizes a rotating dial to adjust the RPM while the spindle is _____, whereas an older belt-driven machine utilizes a step cone pulley system to adjust the RPM while the spindle is _____.

A. in neutral; running
B. turned off; turned off
C. turned on; turned on
D. turned on; turned off

18. Which answer describes a hole drilled 0.750″ deep in a workpiece that has a height of 2.0″?

A. Blind hole
B. Through hole
C. Bottom hole
D. Base hole

19. Which of the following is not recommended when power tapping on a vertical milling machine?

A. Hold the tap in a drill chuck.
B. Apply an abundant amount of cutting oil.
C. "Jog" the spindle to facilitate chip breaking.
D. Set the quill stop to achieve the desired thread depth.

20. What is the purpose of using a soft-faced hammer to tap the top surface of a workpiece that is being held in a vise on a vertical mill?

A. To check the gripping force of the vise
B. To remove any "high spots" from the part's surface
C. To ensure the workpiece is seated evenly against the parallels
D. None of the above

21. The GD&T symbol ∨ represents:

    A. countersink.

    B. chamfer.

    C. counterbore.

    D. angle.

22. While performing a side milling operation, the endmill begins to vibrate and "chatter." Which of the possible solutions listed here would help to solve this problem?

    A. Adjust the workholding so only a minimum amount of material is extending from the vise jaws

    B. Shorten the length of the tool extending from the R-8 collet

    C. Decrease the RPM by 10%

    D. All of the above

    E. None of the above

23. To align the centerline of the spindle with the left edge of the workpiece, how far and in what direction should the spindle be moved from its current position (the tip Ø of the edge finder is 0.200")?

    A. 0.200"; X+

    B. 0.200"; X−

    C. 0.100"; X+

    D. 0.100"; X−

24. When working from a layout to produce a part on a vertical milling machine, which of the tools listed here will locate the center of the spindle over a punch mark on the workpiece?

    A. Wiggler

    B. Cone-pointed edge finder

    C. Centerdrill

    D. All of the above

    E. None of the above

25. A machinist is producing a part on a vertical milling machine. The print specifies a 3/8"-16 UNC-2B thread tapped to a depth of 0.700" ± 0.015". To ensure the thread depth is held to the tolerance, the machinist will need to calculate the distance the tap will advance into the hole per one revolution. What is this amount?

    A. 0.375"

    B. 0.0375"

    C. 0.0234"

    D. 0.0625"

26. Which technique will most likely lead to damage to the endmill and/or the machine?

    A. Calculating the recommended RPM and setting the machine accordingly

    B. Calculating the recommended IPM and setting the machine accordingly

    C. Using cutting oil when applicable

    D. Feeding into the work at a high rate

27. A cast iron part requires several clearance holes and counterbores to be machined at varying locations. Which of the following techniques will assist with these operations?

    A. Feed at a fast rate

    B. Grind the tip of the drill flat

    C. Do not use cutting oil

    D. None of the above

28. Which of the following adjustments should occur only when the spindle on a vertical milling machine is **not** running?

    A. Moving the gear change lever from high to low

    B. Adjusting the variable spindle speed control

    C. Repositioning the feed reversing knob

    D. Engaging the feed control lever

    E. None of the above

29. Which of the following choices is a necessary step in setting up a vertical mill to bore a hole perpendicular to the top of a workpiece within 0.002"?

    A. Align the spindle perpendicular to the table within 0.001"

    B. Align the vise parallel to the spindle

    C. Adjust the ram to line up the spindle with the middle of the table

    D. None of the above

30. The included tip angle measurement of a standard twist drill is:

    A. 118°.

    B. 100°.

    C. 90°.

    D. 82°.

31. Which milling machine component is identified by number 1?

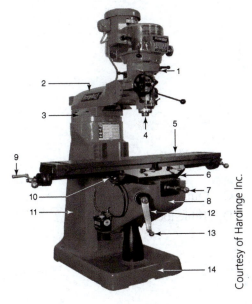

Courtesy of Hardinge Inc.

    A. Ram

    B. Head

    C. Saddle

    D. Chuck

32. Which milling machine component is identified by number 3?

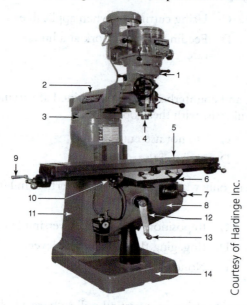

Courtesy of Hardinge Inc.

A. Turret   C. Quill
B. Ram   D. Knee

33. Which milling machine component is identified by number 9?

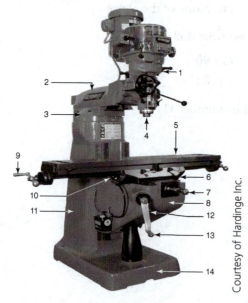

Courtesy of Hardinge Inc.

A. Saddle   C. Quill lock
B. Table crank   D. Table lock

34. Which milling machine component is identified by number 4?

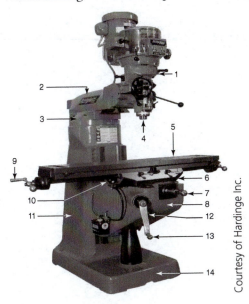

Courtesy of Hardinge Inc.

A. Quill

B. Table

C. Knee

D. Base

35. Which number identifies the component that supports the table?

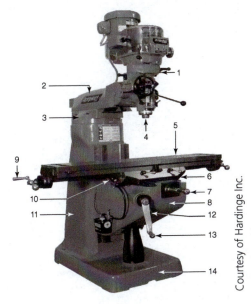

Courtesy of Hardinge Inc.

A. 6

B. 8

C. 10

D. 11

36. Which number identifies the column?

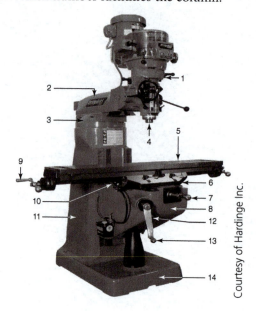

A. 3

B. 8

C. 11

D. 14

37. Which number identifies the knee?

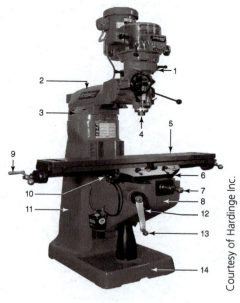

A. 4

B. 6

C. 8

D. 14

38. Which number identifies the table?

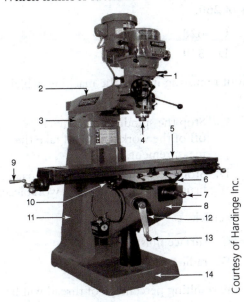

Courtesy of Hardinge Inc.

A. 4
B. 5

C. 6
D. 7

39. Which tool pictured here is a precision layout tool?

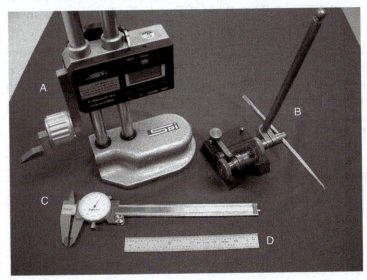

A. A
B. B

C. C
D. D

40. Using 3.82 as the constant, calculate the RPM for a 7/16″ Ø, four-flute endmill machining 1050 plain carbon steel, with a Brinell hardness of 260.

    A. 611

    B. 117

    C. 628

    D. 349

41. To accurately measure the length of a part without removing it from the machine, which sequence of steps must the machinist take?

    A. Stop the spindle, measure, debur, and brush the chips away.

    B. Stop the spindle, remove any burrs, brush the chips away from the part, and take the measurement.

    C. Stop the spindle, blow the chips off of the workpiece, and take the measurement.

    D. Stop the spindle, measure the part, and brush the chips off of the workpiece.

42. A comparison gage and a profilometer are used to check:

    A. squareness.

    B. taper.

    C. surface finish.

    D. radii.

43. When using a tap drill chart to select a tap drill, the resulting percentage of thread will be approximately:

    A. 100%.

    B. 90%.

    C. 75%.

    D. 65%.

44. A part print specifies a slot to be machined on the workpiece at 0.875″ ± 0.003″ wide. The depth of the slot is specified at 0.375″ ± 0.003″. Which method would be the **best** choice for machining this slot?

    A. One pass with a four-flute, 7/8″ Ø endmill, at 3/8″ depth

    B. One roughing pass with a 3/4″ Ø endmill, at 0.360″ depth, and then multiple finishing passes with a 0.500″ Ø endmill, at 0.375″ depth

    C. Two passes with a 0.4375″ Ø endmill, at 3/8″ depth

    D. Four passes with a 7/32″ Ø endmill, at 0.375″ depth

45. When viewing the spindle's rotational direction from above on a vertical milling machine, a right-hand cutting tool will rotate in which direction?

    A. Tangential

    B. Clockwise

    C. X positive

    D. Counterclockwise

46. What considerations must a machinist follow when disposing of solvents, cleaning fluids, and lubricants?

    A. Wear proper PPE.

    B. Reference the MSDS for disposal requirements.

    C. Be prepared with spill prevention and control materials.

    D. All of the above must be followed.

47. An enlarged hole with a flat bottom cut into an existing hole to allow a screw head or nut to sit flush or below a part's surface describes a:

    A. countersink.   C. counterbore.

    B. spotface.   D. clearance hole.

48. A 59/64″ Ø tap drill is the recommended tap drill for which of the following tap sizes?

    A. 1-12 UNF; 3/4-14 NPT   C. M27 × 3; 3/4-14 NPT

    B. 1-8 UNC; 3/8-18 NPT   D. None of the above

49. Which letter in the following image identifies the primary clearance angle of an endmill?

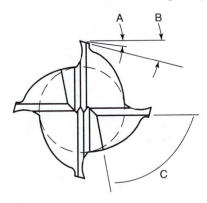

    A. A   C. C

    B. B   D. None of the above

50. When facing a vertical milling machine, the Y-axis moves:

    A. left to right.   C. up and down.

    B. in and out.   D. with the quill feed handle.

5.2 ANSWER KEY

1.	A	26.	D
2.	B	27.	C
3.	B	28.	A
4.	C	29.	A
5.	E	30.	A
6.	B	31.	B
7.	D	32.	A
8.	A	33.	B
9.	B	34.	A
10.	D	35.	A
11.	A	36.	C
12.	D	37.	C
13.	B	38.	B
14.	C	39.	A
15.	C	40.	A
16.	D	41.	B
17.	D	42.	C
18.	A	43.	C
19.	D	44.	B
20.	C	45.	B
21.	A	46.	D
22.	D	47.	C
23.	C	48.	A
24.	D	49.	A
25.	D	50.	B

5.3 PRACTICE THEORY EXAM EXPLANATIONS

1. Which tool could be used to measure the width of a 0.500″ slot milled in a workpiece?

    A. Dial calipers

    B. Depth micrometer

    C. 0″–1″ micrometer

    D. Anvil micrometer

    Answer A is correct. The inside measuring contacts are capable of measuring this part feature.

2. The chip load per tooth for a 1/2″-Ø, two-flute endmill is 0.003″. What is the calculated feed rate in inches per revolution to plunge cut with this endmill?

    A. 0.003″

    B. 0.006″

    C. 0.012″

    D. 0.024″

    Answer B is correct. Multiply the chip load per tooth by the number of cutting edges on the tool to determine the recommended feed increment per revolution of the spindle.

3. Which of the following scenarios could result in breaking the tap?

    A. Prior to tapping, inspect the tap for worn, dull, or chipped cutting edges.

    B. If the recommended tap drill is unavailable, use a smaller-size drill.

    C. Chamfer the hole prior to tapping.

    D. Apply a generous amount of cutting oil.

    E. None of the above could result in breaking the tap.

    Answer B is correct. Using a smaller drill will leave excessive material in the hole, requiring greater torque for the tap to cut. This could exceed the structural integrity of the tap, causing it to break.

4. The micrometer collars on the X-, Y-, and Z- (knee) axes of a vertical milling machine are graduated in _____ increments.

    A. 0.010″

    B. 0.005″

    C. 0.001″

    D. 0.0005″

    Answer C is correct. The micrometer collars on the axes of an inch-based manual milling machine are graduated in 0.001″ increments.

5. Which of the following answers is part of the process for aligning a milling vise parallel to the X-axis on a vertical milling machine?

    A. Assure the vise and table are clean and burr free.

    B. Run a dial indicator across the moveable jaw.

    C. Run a dial indicator across the solid jaw.

    D. Tap the vise into alignment with a soft-face hammer.

    E. A, C, and D are correct.

    Answer E is correct. The process of aligning a vise parallel to the X-axis would include cleaning the vise and table, verifying parallelism with a dial indicator, and fine-tuning the vise alignment with a soft hammer.

6. A machinist is required to machine the angle specified on the following part print. Which setup method would be applicable for this operation?

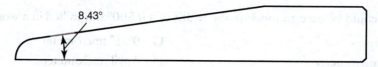

A. Hold the workpiece in a vise and tilt the head to the required angle.

B. Use a sine bar to set the angle and clamp the part to an angle block along the length of the workpiece.

C. Clamp the piece in a vise with the angled surface extending from the side of the vise and use the graduations on the swivel base to set the angle.

D. None of the above would be applicable for this operation.

Answer B is correct. Using gage blocks, the sine bar can be set to a precise angle. The workpiece can then be positioned against the angle plate and secured with clamps at the specified angle.

7. The _____ is a component on a vertical milling machine that pulls the tool holder into the R-8 taper located on the inside of the spindle.

A. retention rod

B. spindle clamp

C. pull stud

D. drawbar

Answer D is correct. A drawbar is a steel rod with external threads cut on one end and a hex nut on the other. The drawbar is installed through the spindle of the milling machine. As the drawbar is tightened, the tool holder is pulled into the R-8 taper, securing the tool in the spindle.

8. The process of aligning the head on a vertical milling machine perpendicular to the table in the *X*- and *Y*-axes is called:

A. tramming.

B. squaring.

C. milling.

D. quill locating.

Answer A is correct. To ensure part geometry is produced parallel and perpendicular to the machine's table, the head must be aligned to the table's top surface. This alignment process is called tramming.

9. To align the vertical centerline of the woodruff keyseat cutter with the vertical centerline of the shaft in the following image, what adjustment to the machine is necessary? The shaft diameter is 1.0″ and the woodruff cutter is 0.250″ thick.

A. The table should be raised 0.500″.

B. The table should be raised 0.500″ + 0.125″.

C. The table should be raised 0.375″.

D. The table should be raised 0.500″ + 0.250″.

Answer B is correct. The cutting tool is touched off the top of the 1.0″-diameter material. Raising the workpiece 0.500″ will align the bottom of the cutter to the centerline of the shaft. Raising it an additional 0.125″ will locate the vertical centerlines at the same position.

10. The print specifies 5/16″ radii in the four corners of a manually milled pocket. What size endmill will be necessary when finishing machining this pocket to produce the specified radii?

A. 0.250″ Ø

B. 0.3125″ Ø

C. 0.4375″ Ø

D. 0.625″ Ø

Answer D is correct. When manually milling pockets, the corner radii will be determined by the radius of the endmill. An endmill diameter of 0.625″ will leave a corner radius of 0.3125″.

11. A machinist is producing a part on a vertical milling machine with no power feed. Which of the following methods is considered safe and acceptable?

A. Conventional milling during roughing operations

B. Extending the endmill from the collet 1/2″ past the top of the cutting edges

C. Using a feed rate that is twice the recommended rate to save machining time

D. Climb milling for all side milling operations

Answer A is correct. Conventional milling is considered a safe method because the tool is fed against the rotational direction of the cutter and will push away from the work rather than trying to pull the tool into the work.

12. Which of the following variables has the greatest effect on determining the RPM for a milling operation?

   A. The number of flutes on the cutter
   B. The horsepower of the machine
   C. The tolerance of the part features
   D. The material being machined

   Answer D is correct. The machinability or measure of how easily a material is capable of being machined will vary between different types of material. Harder materials have characteristics that make machining more challenging than softer materials, therefore requiring different cutting RPM.

13. A part print specifies a hole diameter as 19/64″ (0.2969) ± 0.001″. Which combination of gage pins could be used as a go/no-go gage for this dimension?

   A. 0.295″ and 0.299″
   B. 0.296″ and 0.298″
   C. 0.295″ and 0.298″
   D. None of these

   Answer B is correct. The upper limit of this feature is 0.2979″ and the lower limit is 0.2959″. The 0.296″ gage pin would be the go pin and the 0.298″ pin would be the no-go pin.

14. To machine hole "A" on the following part print and hold the location specifications, which method would be most accurate?

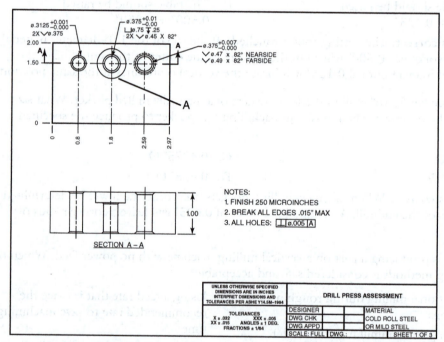

   A. Lay out the hole locations with a height gage, center punch, use a hand drill to centerdrill, and drill.

   B. Lay out the hole locations with a height gage, center punch, set up on a drill press, use a center finder to locate the position, centerdrill, and drill.

   C. Set up the workpiece on a vertical mill, locate the workpiece origin with an edge finder, use the DRO to position the table at the hole location, centerdrill, and drill.

   D. Lay out the hole location, set up the piece on a vertical milling machine, use a cone-pointed edge finder to locate the position, spot drill, and drill.

   Answer C is correct. Using an edge finder with a square workpiece and properly trammed milling machine head will provide a setup capable of holding these location specifications.

15. What type of tap should be used to produce the thread to the required depth on hole "B"?

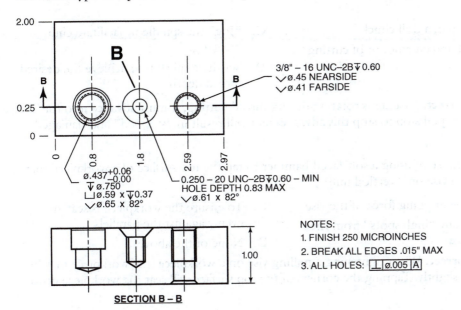

3/8" – 16 UNC–2B ▼ 0.60
⌄ø.45 NEARSIDE
⌄ø.41 FARSIDE

ø.437 +0.06 −0.00
▼ø.750
⊔ø.59 x ▼0.37
⌄ø.65 x 82°

0.250 – 20 UNC–2B▼0.60 – MIN
HOLE DEPTH 0.83 MAX
⌄ø.61 x 82°

NOTES:
1. FINISH 250 MICROINCHES
2. BREAK ALL EDGES .015" MAX
3. ALL HOLES: ⟂ ø.005 A

SECTION B – B

A. Taper
B. Plug
C. Bottoming
D. Spiral

Answer C is correct. A bottoming tap has a small tip chamfer on only the first one/two threads. This allows the tap to produce full threads to just about the entire hole depth.

16. The standard taper inside the spindle of a vertical milling machine, which is used to center the tool holder, is known as a:

A. 5C.
B. 29°.
C. Morse.
D. R-8.

Answer D is correct. Most manual vertical milling machines are equipped with a standard R-8 tapper.

17. A modern milling machine equipped with a variable speed control utilizes a rotating dial to adjust the RPM while the spindle is _____, whereas an older belt-driven machine utilizes a step cone pulley system to adjust the RPM while the spindle is _____.

A. in neutral; running
B. turned off; turned off
C. turned on; turned on
D. turned on; turned off

Answer D is correct. The speed control mechanisms on a variable-speed milling machine require the spindle to be on when changing the RPM. The older belt-driven machines require the belt to be physically moved from one pulley to another, which is done only when the power is off.

18. Which answer describes a hole drilled 0.750″ deep in a workpiece that has a height of 2.0″?

A. Blind hole
B. Through hole
C. Bottom hole
D. Base hole

Answer A is correct. A blind hole is one that is drilled to a depth or only partially through a workpiece.

19. Which of the following is not recommended when power tapping on a vertical milling machine?

    A. Hold the tap in a drill chuck.

    B. Apply an abundant amount of cutting oil.

    C. "Jog" the spindle to facilitate chip breaking.

    D. Set the quill stop to achieve the desired thread depth.

    Answer D is correct. As a tap is rotating during thread cutting, it must advance into the hole. Setting the quill stop to stop this advancement will result in "stripped" and damaged threads.

20. What is the purpose of using a soft-faced hammer to tap the top surface of a workpiece that is being held in a vise on a vertical mill?

    A. To check the gripping force of the vise

    B. To remove any "high spots" from the part's surface

    C. To ensure the workpiece is seated evenly against the parallels

    D. None of the above

    Answer C is correct. When tightening a milling vise on a workpiece, the workpiece may lift off the parallels slightly. Tapping the corners of the top surface will seat the workpiece down on the parallels.

21. The GD&T symbol $\vee$ represents:

    A. countersink.

    B. chamfer.

    C. counterbore.

    D. angle.

    Answer A is correct. The Geometric Dimensioning and Tolerancing symbol for a countersink is $\vee$.

22. While performing a side milling operation, the endmill begins to vibrate and "chatter." Which of the possible solutions listed here would help to solve this problem?

    A. Adjust the workholding so only a minimum amount of material is extending from the vise jaws

    B. Shorten the length of the tool extending from the R-8 collet

    C. Decrease the RPM by 10%

    D. All of the above

    E. None of the above

    Answer D is correct. Several controllable variables affect the cutting conditions of a milling operation. A ridged setup of both the workpiece and the tool, combined with the optimal RPM, will minimize cutting tool vibration.

23. To align the centerline of the spindle with the left edge of the workpiece, how far and in what direction should the spindle be moved from its current position (the tip Ø of the edge finder is 0.200″)?

   A. 0.200″; X+   C. 0.100″; X+

   B. 0.200″; X−   D. 0.100″; X−

Answer C is correct. The image shows that the tip of the edge finder has just "kicked out," indicating the side of the tip is directly adjacent to the workpiece. To align the centerline of the spindle with the edge of the workpiece, the spindle would need to move in the X+ direction, one half the diameter of the edge-finder tip.

24. When working from a layout to produce a part on a vertical milling machine, which of the tools listed here will locate the center of the spindle over a punch mark on the workpiece?

   A. Wiggler   D. All of the above

   B. Cone-pointed edge finder   E. None of the above

   C. Centerdrill

Answer D is correct. A wiggler, cone-pointed edge finder, and centerdrill are all tools that could be used to locate the position of a punch mark.

25. A machinist is producing a part on a vertical milling machine. The print specifies a 3/8″−16 UNC-2B thread tapped to a depth of 0.700″ ± 0.015″. To ensure the thread depth is held to the tolerance, the machinist will need to calculate the distance the tap will advance into the hole per one revolution. What is this amount?

   A. 0.375″   C. 0.0234″

   B. 0.0375″   D. 0.0625″

Answer D is correct. The pitch of a screw thread can be calculated by dividing 1 by the number of threads per inch (Pitch = 1/16).

26. Which technique will most likely lead to damage to the endmill and/or the machine?

    A. Calculating the recommended RPM and setting the machine accordingly

    B. Calculating the recommended IPM and setting the machine accordingly

    C. Using cutting oil when applicable

    D. Feeding into the work at a high rate

    Answer D is correct. Excessive feed rates can lead to accelerated tool wear and damage.

27. A cast iron part requires several clearance holes and counterbores to be machined at varying locations. Which of the following techniques will assist with these operations?

    A. Feed at a fast rate

    B. Grind the tip of the drill flat

    C. Do not use cutting oil

    D. None of the above

    Answer C is correct. The internal properties and heat conductivity of cast iron all but eliminate any advantage of using cutting oil.

28. Which of the following adjustments should occur only when the spindle on a vertical milling machine is **not** running?

    A. Moving the gear change lever from high to low

    B. Adjusting the variable spindle speed control

    C. Repositioning the feed reversing knob

    D. Engaging the feed control lever

    E. None of the above

    Answer A is correct. The high-/low-range lever controls the movement of gears within the head's speed transmission. Attempting to move this lever with the machine turned on will result in damage to this transmission.

29. Which of the following choices is a necessary step in setting up a vertical mill to bore a hole perpendicular to the top of a workpiece within 0.002"?

    A. Align the spindle perpendicular to the table within 0.001"

    B. Align the vise parallel to the spindle

    C. Adjust the ram to line up the spindle with the middle of the table

    D. None of the above

    Answer A is correct. If the perpendicularity specification is going to be held, the spindle must be aligned perpendicular to the workpiece within this amount (0.002").

30. The included tip angle measurement of a standard twist drill is:

    A. 118°.

    B. 100°.

    C. 90°.

    D. 82°.

    Answer A is correct. For most materials, 118° is the recommended point angle for general drilling operations.

31. Which milling machine component is identified by number 1?

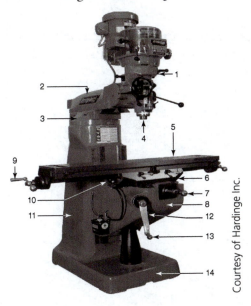

Courtesy of Hardinge Inc.

A. Ram   C. Saddle

B. Head   D. Chuck

Answer B is correct. Number 1 is pointing to the head.

32. Which milling machine component is identified by number 3?

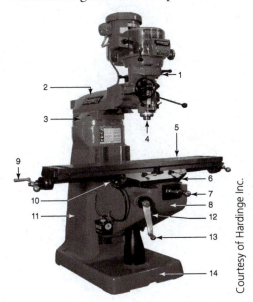

Courtesy of Hardinge Inc.

A. Turret   C. Quill

B. Ram   D. Knee

Answer A is correct. Number 3 is pointing to the turret.

33. Which milling machine component is identified by number 9?

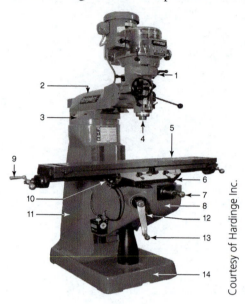

Courtesy of Hardinge Inc.

A. Saddle   C. Quill lock
B. Table crank   D. Table lock

Answer B is correct. Number 9 is pointing to the table crank.

34. Which milling machine component is identified by number 4?

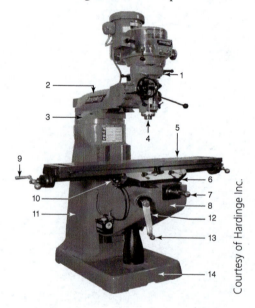

Courtesy of Hardinge Inc.

A. Quill   C. Knee
B. Table   D. Base

Answer A is correct. Number 4 is pointing to the quill.

35. Which number identifies the component that supports the table?

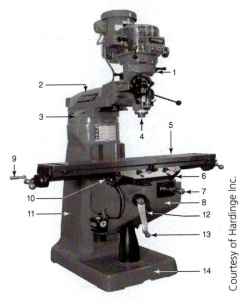

Courtesy of Hardinge Inc.

E. 6   G. 10
F. 8   H. 11

Answer A is correct. Number 6 is pointing to the saddle that supports the table.

36. Which number identifies the column?

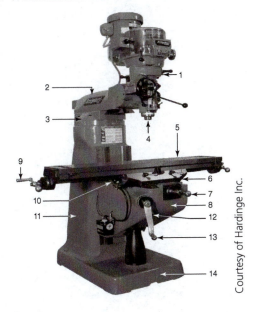

Courtesy of Hardinge Inc.

A. 3   C. 11
B. 8   D. 14

Answer C is correct. Number 11 is pointing to the column.

37. Which number identifies the knee?

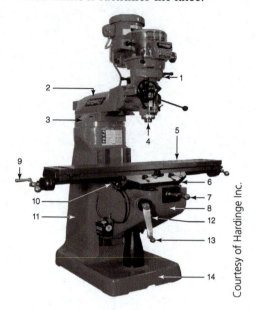

A. 4

B. 6

C. 8

D. 14

Answer C is correct. Number 8 is pointing to the knee.

38. Which number identifies the table?

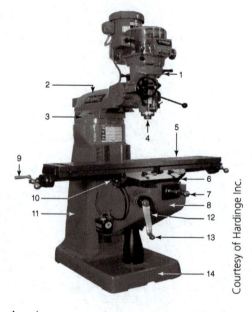

A. 4

B. 5

C. 6

D. 7

Answer B is correct. Number 5 is pointing to the table.

39. Which tool pictured here is a precision layout tool?

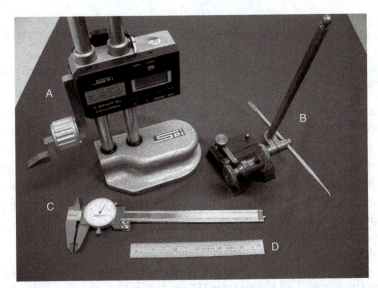

A. A

B. B

C. C

D. D

Answer A is correct. Letter A identifies a digital height gage, which when used with a surface plate is capable of producing precise layout lines.

40. Using 3.82 as the constant, calculate the RPM for a 7/16″ Ø, four-flute endmill machining 1050 plain carbon steel, with a Brinell hardness of 260.

A. 611

B. 117

C. 628

D. 349

Answer A is correct. Inserting the provided variables into the RPM formula will result in

$$RPM = \frac{3.82 \times 70}{.4375} = 611.2$$

41. To accurately measure the length of a part without removing it from the machine, which sequence of steps must the machinist take?

A. Stop the spindle, measure, debur, and brush the chips away.

B. Stop the spindle, remove any burrs, brush the chips away from the part, and take the measurement.

C. Stop the spindle, blow the chips off of the workpiece, and take the measurement.

D. Stop the spindle, measure the part, and brush the chips off of the workpiece.

Answer B is correct. The spindle must be stopped prior to measuring to prevent serious injuries from occurring, and any burrs must be removed so as not to interfere with the measurement. After deburring, any chips should be safely removed with a brush.

42. A comparison gage and a profilometer are used to check:

A. squareness.

B. taper.

C. surface finish.

D. radii.

Answer C is correct. Surface finish can be visually inspected by comparing the workpiece to a comparison gage, and profilometer can be used for a more accurate measurement of surface roughness.

43. When using a tap drill chart to select a tap drill, the resulting percentage of thread will be approximately:

    A. 100%.   C. 75%.
    B. 90%.   D. 65%.

    Answer C is correct. The majority of tap drill charts recommend tap drill sizes that provide a thread percentage ranging from 51% to 80%, with an average of 75%.

44. A part print specifies a slot to be machined on the workpiece at 0.875″ ± 0.003″ wide. The depth of the slot is specified at 0.375″ ± 0.003″. Which method would be the **best** choice for machining this slot?

    A. One pass with a four-flute, 7/8″ Ø endmill, at 3/8″ depth

    B. One roughing pass with a 3/4″ Ø endmill, at 0.360″ depth, and then multiple finishing passes with a 0.500″ Ø endmill, at 0.375″ depth

    C. Two passes with a 0.4375″ Ø endmill, at 3/8″ depth

    D. Four passes with a 7/32″ Ø endmill, at 0.375″ depth

    Answer B is correct. A roughing pass along the centerline of the slot with a 3/4″ Ø endmill will remove the majority of the material and leave 0.0625″ on each side of the slot width for finishing. The 0.500″ Ø endmill can then be used to make successive passes to machine the slot to the specified width, depth, and location.

45. When viewing the spindle's rotational direction from above on a vertical milling machine, a right-hand cutting tool will rotate in which direction?

    A. Tangential   C. X positive
    B. Clockwise   D. Counterclockwise

    Answer B is correct. To cut properly, a right-hand cutting tool will rotate in the clockwise direction when viewed from above.

46. What considerations must a machinist follow when disposing of solvents, cleaning fluids, and lubricants?

    A. Wear proper PPE.   C. Be prepared with spill prevention and control materials.
    B. Reference the MSDS for disposal requirements.
                                      D. All of the above must be followed.

    Answer D is correct. When handling and disposing of chemicals, fluids, and lubricants, the corresponding MSDS should be referenced for the required PPE and cleanup methods.

47. An enlarged hole with a flat bottom cut into an existing hole to allow a screw head or nut to sit flush or below a part's surface describes a:

    A. countersink.   C. counterbore.
    B. spotface.   D. clearance hole.

    Answer C is correct. A counterbore is a cylindrical enlargement cut into an existing hole. The purpose of a counterbore is often to allow a nut or the head of a fastener to sit flush or below a part's surface.

48. A 59/64″ Ø tap drill is the recommended tap drill for which of the following tap sizes?

    A. 1-12 UNF; 3/4-14 NPT   C. M27 × 3; 3/4-14 NPT
    B. 1-8 UNC; 3/8-18 NPT   D. None of the above

    Answer A is correct. Referencing a Unified National Series Tap Drill chart, a 59/64″ Ø tap drill is specified for both a 1–12 UNF and a 3/4-14 NPT tap.

49. Which letter in the following image identifies the primary clearance angle of an endmill?

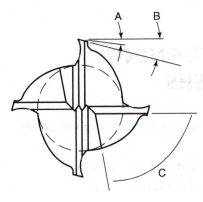

    A. A   C. C
    B. B   D. None of the above

    Answer A is correct. The primary side clearance angle is the relief angle adjacent to the cutting edge and is typically ground between 5° and 9°.

50. When facing a vertical milling machine, the Υ-axis moves:

    A. left to right.   C. up and down.
    B. in and out.   D. with the quill feed handle.

    Answer B is correct. The axes of a vertical milling machine are identified by the Cartesian coordinate system. When facing the machine, the Υ-axis moves in the in/out direction.

6 Turning Operations Certifications

6.1 PRACTICE THEORY EXAM—TURNING CHUCKING/BETWEEN CENTERS

1. A machinist is calculating the spindle speed for a turning operation; the diameter variable in the RPM formula is determined by:

   A. the size of the chuck.

   B. the radius on the cutting tool.

   C. the diameter of the workpiece.

   D. None of the above

2. An 8.75"-long shaft is being turned at 560 RPM, using a feed rate of 0.005" per revolution. How many minutes will it take to make one pass?

   A. 3.25

   B. 2.513

   C. 2.125

   D. 3.125

3. Choose the statement that **best** defines the cutting speed on a lathe.

   A. A rate measured in surface feet per minute (SFM)

   B. The ratio of machine horsepower to workpiece diameter

   C. The distance a point on the circumference of a rotating workpiece travels past the tool in one minute

   D. Both A and C

4. A machinist is setting up a lathe to turn the 1.975" Ø OD on the part in the following figure. Using 3.82 as the constant, what RPM should the machine be set to?

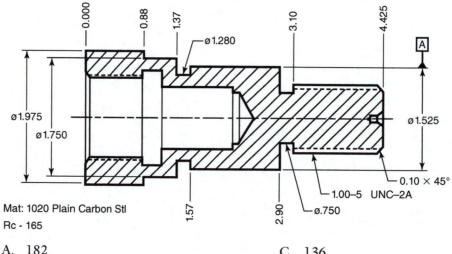

Mat: 1020 Plain Carbon Stl
Rc - 165

   A. 182

   B. 313

   C. 136

   D. 213

5. As part of the job planning process, a machinist is trying to estimate the time required to perform the turning operations on a part. Which formula would be correct for this calculation? Note: L = length of cut, T = time, N = spindle speed (RPM), f = feed (IPR)

A. $T = L \times f \times N$

B. $T = L \times f \div N$

C. $T = L \div (f \times N)$

D. $T = (N \times f) \div L$

6. Which of the following situations would **not** present a potential safety hazard when operating a lathe?

A. Turning the machine off and removing stringy chips with a chip hook

B. Using a dead center without lube

C. Temporarily removing a guard

D. Exceeding the maximum RPM rating of the lathe chuck

7. Which of the following categories would be found on an MSDS?

A. Recommended PPE, disposal considerations, flash point, chemical characteristics

B. First-aid measures, PPE, price, handling and storage

C. Price, boiling point, expiration date, first-aid measures

D. Potential health effects, handling and storage, CEO of the manufacturing company, odor description

8. Which of the following is **not** a safe practice when using a lathe?

A. Never reach across a rotating chuck or workpiece.

B. Use a brush to clean the lathe and work area and then sweep up the work area around the lathe.

C. Check that the workpiece is secured in the chuck and remove the chuck key prior to starting the lathe.

D. Wear a long-sleeved shirt so it covers the watch you are wearing.

9. Which tool is used for aligning the threading tool to a workpiece and checking the angle of the threading tool?

A. Center gage

B. Pitch micrometer

C. Protractor

D. Screw pitch gage

10. You are setting up a lathe to cut a right-hand 1/2-13 UNC-2A thread. The compound rest is currently positioned perpendicular to the workpiece. The compound rest will need to be positioned:

A. 29 1/2° to the left.

B. 60° to the left.

C. 60° to the right.

D. 29 1/2° to the right.

11. Use the formula Compound infeed = 0.7 × pitch and calculate the amount of compound rest infeed for a 1/2-20 UNF-2A.

A. 1.4″

B. 0.0035″

C. 0.035″

D. 0.14″

12. A thread micrometer is used to measure what dimension on an external thread?

A. Major diameter

B. Minor diameter

C. Thread helix angle

D. Pitch diameter

13. The sleeve on an inch-based micrometer has divisions equal to _____ and graduations on the thimble equal to _____.

    A. 0.025"; 0.001"
    B. 0.025"; 0.010"

    C. 0.001"; 0.010"
    D. 0.001"; 0.025"

14. When determining the appropriate measuring device for inspecting a part feature, the machinist must first assess the:

    A. accuracy of the machine.
    B. tolerance of the dimension.

    C. quantity of parts to inspect.
    D. Both A and C.

15. Checking the runout of a headstock center would require which of the following measuring tools?

    A. Telescope gage
    B. Height gage

    C. Dial indicator
    D. Depth micrometer

16. Select the statement that is most accurate.

    A. A lathe center contains a pointed end ground on a 60° angle with an R-8 taper on the other end.
    B. A lathe center is a device that is installed only in a tailstock and is ground to a 30° point.

    C. A lathe center is cylindrical, ground to a point at a 60° angle on one end, and contains a Morse taper on the other end.
    D. A lathe center contains a Morse taper on one end and is ground to a point at a 90° angle on the other end.

17. Select the answer(s) that **best** identifies the result of turning between centers.

    A. Runout is minimized.
    B. Work can be removed and reinstalled with a high level of repeatability.

    C. Dead centers and live centers both contain rotating parts.
    D. All of the above are correct.
    E. Both A and B are correct.

18. Select the answer that is **not** a characteristic of a universal chuck.

    A. The movement of all the jaws is synchronized.
    B. Each jaw is capable of being moved independently.

    C. The jaws are "self-centering" and move in and out by turning a chuck key.
    D. There are two-, three-, or six-piece jaw configurations.

19. Select the workholding device that is defined incorrectly.

    A. Faceplate—mounted to the headstock spindle and is used for irregular-shaped work
    B. Universal chuck—jaws open and close simultaneously

    C. Collet—accurate, repeatable, and quick to interchange parts
    D. Four-jaw chuck—can be used for eccentric or off-center workholding of triangular-shaped workpieces

20. A machinist must remove 0.025" from a 1.500" Ø × 25.00"–long shaft. To safely and accurately complete this task, the setup should include:

    A. a follower rest.
    B. the use of the half-nut lever.

    C. a taper reducer.
    D. a solid mandrel.

21. Which workholding device would include the following operating sequence: Open the jaws slightly larger than the diameter of the stock, number each jaw with a marker, lightly clamp the stock while visibly aligning it to center, and use a dial indicator to check/adjust until the desired TIR is achieved?

    A. Faceplate

    B. Universal chuck

    C. Independent chuck

    D. Step collet

22. The machinist engages the half-nut lever. Which of the following specifications corresponds to this action?

    A. 6-32 UNF-2B

    B. Ø 0.750″ ± 0.005

    C. 1/2″ TPI

    D. 1/2-13 UNC-2A

23. The carriage of a lathe is moving 0.003″ per revolution of the spindle. This movement is determined by:

    A. spindle clutch lever.

    B. quick change gear box.

    C. carriage traverse lever.

    D. step pulley.

24. The cross slide moves in a direction _____ to the face of the part and at a speed _____ to that of the longitudinal feed.

    A. parallel; one-third to one-half

    B. parallel; equal

    C. perpendicular; one-third to one-half

    D. perpendicular; two times

25. The size of a lathe is determined by the _____ and the _____.

    A. horsepower; swing

    B. swing; bed length

    C. length; width

    D. maximum allowable tool size; maximum RPM

26. Which of the following **best** describes the characteristics of a taper?

    A. Measured in degrees, taper per inch, taper per foot

    B. Continuous change in diameter

    C. Can be self-holding or self-releasing

    D. All of the above

27. The RPM for a reaming operation is approximately _____ the RPM of a drilling operation of the same diameter.

    A. one-half

    B. one-fourth

    C. equal to

    D. two times

28. A facing tool is reinstalled after it is sharpened. The first cut produces a **cone-shaped** protrusion at the center of the piece. What is the probable root cause?

    A. The tool is on center.

    B. The tool is below center.

    C. The tool is above center.

    D. The RPM is set too high.

29. A raised pattern on the surface of a workpiece is created by displacing the material with two hardened wheels. The basic patterns created by this process are straight and diamond. Identify this process.

    A. Gripping

    B. Pressing

    C. Knurling

    D. Etching

30. Which of the following statements is true about the initial cut on a 3.0″ cast iron shaft?

 A. A generous amount of cutting oil should be applied.

 B. The depth of cut should penetrate below the rough and hard outer scale.

 C. Reduce the cutting speed.

 D. None of the above.

31. The control charts used with SPC to analyze trends in part variation are:

 A. X bar and R charts.

 B. production and cost charts.

 C. capability and first-piece inspection charts.

 D. X graph and Y charts.

32. An R chart or range chart shows the variation of each sampling. What is the definition of range in this example?

 A. The average size of a dimension in a sampling

 B. The average time it takes to machine a part

 C. A value obtained by subtracting the smallest dimension from the largest dimension in a sample data set

 D. The maximum distance a measuring tool is capable of measuring

33. Select the most reliable method described here for repeatable results when turning the length of a shoulder on the lathe that is not equipped with a DRO.

 A. Set the compound rest parallel to the centerline of the lathe.

 B. Set the compound rest perpendicular to the centerline of the lathe.

 C. Support the piece with a tailstock.

 D. Install a micrometer stop to the right of the tailstock.

34. The upper left-hand area of the lathe identified by letter B is the _____, which houses the _____ and the _____.

 A. motor; belts; drive screw

 B. gears; clutch; sleeve

 C. cabinet; transmission; collar

 D. headstock; gears; spindle

35. Letter C identifies the:

A. apron.

B. table.

C. tool carrier.

D. compound rest.

36. The carriage slides along the _____, which are identified by letter E.

A. rails

B. tracks

C. guides

D. ways

37. Letter F represents _____, the casting that runs the length of the lathe.

A. base

B. bed

C. frame

D. column

38. Letter H identifies the _____, which rotates _____.

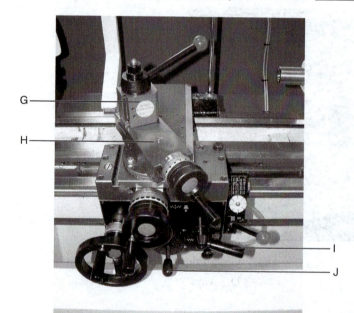

A. compound rest; 360°

B. cross slide; 360°

C. compound rest; 180°

D. tool post; 180°

39. Letter I identifies the:

A. half-nut lever.

B. feed lever.

C. spindle start/stop lever.

D. thread retract lever.

40. Letter J identifies the:

A. half-nut lever.

B. feed lever.

C. spindle start/stop lever.

D. thread retract lever.

41. Letter K identifies the:

   A. measurement dial.
   B. dial indicator.

   C. fine drilling gage.
   D. micrometer collar.

42. Letter L identifies the:

   A. ram.
   B. tailstock piston.

   C. tailstock quill.
   D. subspindle.

43. Letter M identifies the:

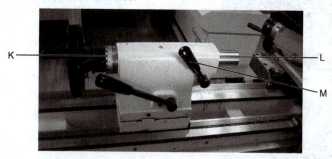

   A. feed lever.
   B. tailstock quill lock.

   C. tailstock retract lever.
   D. half-nut lever.

44. A tailstock that is offset toward the operator will produce a turned workpiece that is:

   A. tapered with the end near the tailstock being smaller.
   B. undersize the entire length.

   C. tapered with the end near the headstock being smaller.
   D. oversize the entire length.

45. Precise alignment of the centers on a lathe while turning a workpiece can be accomplished by:

    A. visually aligning the graduation marks on the base of the tailstock.

    B. using a test bar and adjusting a test indicator as necessary.

    C. taking a trial cut, measuring each end with a micrometer, and adjusting.

    D. bringing the centers together by sliding the tailstock forward, adjusting, and checking visually.

46. Select the acronym that represents the class of fit that is intended to provide similar running capability while providing for lubrication allowance.

    A. FN

    B. RC

    C. LC

    D. LT

47. Refer to *The Machinery's Handbook* to select the applicable shaft diameter for an FN2 force fit with a 2.625″-diameter hole:

    A. 2.6225″.

    B. 2.6253″.

    C. 2.6262″.

    D. 2.6275″.

48. Refer to *The Machinery's Handbook* to select the applicable bore diameter for an RC4 clearance fit with a 1.300″-diameter shaft.

    A. 1.3010″

    B. 1.3030″

    C. 1.2980″

    D. 1.2990″

    E. 0.0016

49. Refer to *The Machinery's Handbook* to determine the appropriate tap drill to produce a tapered 1/4-18 NPT:

    A. 37/64 drill.

    B. 7/32 drill.

    C. 7/16 drill.

    D. 29/64 drill.

50. Which of the following serves as a reference plane when used with a height gage to produce precision layouts?

    A. Surface plate

    B. Angle plate

    C. Surface gage

    D. Protractor

6.2 ANSWER KEY

1.	C		26.	D
2.	D		27.	A
3.	D		28.	C
4.	D		29.	C
5.	C		30.	B
6.	A		31.	A
7.	A		32.	C
8.	D		33.	A
9.	A		34.	D
10.	D		35.	A
11.	C		36.	D
12.	D		37.	B
13.	A		38.	A
14.	B		39.	A
15.	C		40.	B
16.	C		41.	D
17.	E		42.	C
18.	B		43.	B
19.	D		44.	A
20.	A		45.	C
21.	C		46.	B
22.	D		47.	D
23.	B		48.	A
24.	A		49.	C
25.	B		50.	A

6.3 PRACTICE THEORY EXAM EXPLANATIONS

1. A machinist is calculating the spindle speed for a turning operation; the diameter variable in the RPM formula is determined by:

    A. the size of the chuck.
    B. the radius on the cutting tool.
    C. the diameter of the workpiece.
    D. None of the above

    Answer C is correct. The diameter of the workpiece is directly proportional to the circumference that determines the amount of material that will pass by the tool in a minute's time.

2. An 8.75″-long shaft is being turned at 560 RPM, using a feed rate of 0.005″ per revolution. How many minutes will it take to make one pass?

    A. 3.25
    B. 2.513
    C. 2.125
    D. 3.125

    Answer D is correct. Use the following formula:
    Time (minutes) = $L \div$ RPM $\times$ *feed rate* $3.125 = 8.75 \div 560 \times 0.005$.

3. Choose the statement that **best** defines the cutting speed on a lathe.

    A. A rate measured in surface feet per minute (SFM)
    B. The ratio of machine horsepower to workpiece diameter
    C. The distance a point on the circumference of a rotating workpiece travels past the tool in one minute
    D. Both A and C

    Answer D is correct. The combination of both answers A and C is the *best* definition.

4. A machinist is setting up a lathe to turn the 1.975″ Ø OD on the part in the following figure. Using 3.82 as the constant, what RPM should the machine be set to?

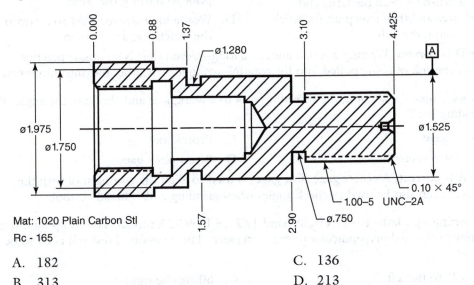

Mat: 1020 Plain Carbon Stl
Rc - 165

    A. 182
    B. 313
    C. 136
    D. 213

    Answer D is correct. Use the formula RPM = CS × 3.82 ÷ Ø. The value for CS is 110, which is obtained from the chart in *The Machinery's Handbook*. The calculations should equal 212.759, which rounds to 213.

5. As part of the job planning process, a machinist is trying to estimate the time required to perform the turning operations on a part. Which formula would be correct for this calculation? Note: L = length of cut, T = time, N = spindle speed (RPM), f = feed (IPR)

   A. $T = L \times f \times N$ C. $T = L \div (f \times N)$

   B. $T = L \times f \div N$ D. $T = (N \times f) \div L$

   Answer C is correct. $T = L \div (f \times N)$ is the correct formula for calculating machining time.

6. Which of the following situations would **not** present a potential safety hazard when operating a lathe?

   A. Turning the machine off and removing stringy chips with a chip hook   C. Temporarily removing a guard

   B. Using a dead center without lube   D. Exceeding the maximum RPM rating of the lathe chuck

   Answer A is correct. Stringy metal chips should be removed with a chip hook or pliers after the machine is stopped.

7. Which of the following categories would be found on an MSDS?

   A. Recommended PPE, disposal considerations, flash point, chemical characteristics   C. Price, boiling point, expiration date, first-aid measures

   B. First-aid measures, PPE, price, handling and storage   D. Potential health effects, handling and storage, CEO of the manufacturing company, odor description

   Answer A is correct. Personal protective equipment recommendations, disposal considerations, flash point, and chemical information and ingredients are categories found on an MSDS.

8. Which of the following is **not** a safe practice when using a lathe?

   A. Never reach across a rotating chuck or workpiece.   C. Check that the workpiece is secured in the chuck and remove the chuck key prior to starting the lathe.

   B. Use a brush to clean the lathe and work area and then sweep up the work area around the lathe.   D. Wear a long-sleeved shirt so it covers the watch you are wearing.

   Answer D is correct. Wearing a watch and/or a long-sleeved shirt is an unsafe practice because of the risk of being pulled into the rotating workpiece or other moving components.

9. Which tool is used for aligning the threading tool to a workpiece and checking the angle of the threading tool?

   A. Center gage   C. Protractor

   B. Pitch micrometer   D. Screw pitch gage

   Answer A is correct. A center gage (fish gage) is used to align the cutting tool with the workpiece. It can also be used to check angles when grinding a 60° threading tool.

10. You are setting up a lathe to cut a right-hand 1/2-13 UNC-2A thread. The compound rest is currently positioned perpendicular to the workpiece. The compound rest will need to be positioned:

   A. 29 1/2° to the left.   C. 60° to the right.

   B. 60° to the left.   D. 29 1/2° to the right.

   Answer D is correct. 30°to the right is the bisector of the 60° included thread angle. By setting the compound rest to 29 1/2°, you will allow the right side of the cutting tool to take a small skim cut with each pass.

11. Use the formula Compound infeed = 0.7 × pitch and calculate the amount of compound rest infeed for a 1/2-20 UNF-2A.

   A. 1.4″

   B. 0.0035″

   C. 0.035″

   D. 0.14″

   Answer C is correct. Inserting the provided data into the given formula would look like this: Compound infeed = 0.7 × 0.05 (*Hint: pitch = 1/number of threads per inch*). This will result in an answer of 0.035″.

12. A thread micrometer is used to measure what dimension on an external thread?

   A. Major diameter

   B. Minor diameter

   C. Thread helix angle

   D. Pitch diameter

   Answer D is correct. The screw thread micrometer is the applicable tool for measuring the pitch diameter of an external thread.

13. The sleeve on an inch-based micrometer has divisions equal to _____ and graduations on the thimble equal to _____.

   A. 0.025″; 0.001″

   B. 0.025″; 0.010″

   C. 0.001″; 0.010″

   D. 0.001″; 0.025″

   Answer A is correct. The divisions on an inch-based micrometer sleeve are 0.025″ each, and the divisions or graduations on the thimble are 0.001″ each.

14. When determining the appropriate measuring device for inspecting a part feature, the machinist must first assess the:

   A. accuracy of the machine.

   B. tolerance of the dimension.

   C. quantity of parts to inspect.

   D. Both A and C.

   Answer B is correct. The tolerance of a dimension to be inspected governs whether to use a semiprecision or a precision measuring tool for checking that dimension.

15. Checking the runout of a headstock center would require which of the following measuring tools?

   A. Telescope gage

   B. Height gage

   C. Dial indicator

   D. Depth micrometer

   Answer C is correct. A dial indicator is used to check surface deviation and would be the applicable measuring tool for this task.

16. Select the statement that is most accurate.

   A. A lathe center contains a pointed end ground on a 60° angle with an R-8 taper on the other end.

   B. A lathe center is a device that is installed only in a tailstock and is ground to a 30° point.

   C. A lathe center is cylindrical, ground to a point at a 60° angle on one end, and contains a Morse taper on the other end.

   D. A lathe center contains a Morse taper on one end and is ground to a point at a 90° angle on the other end.

   Answer C is correct. A lathe center is cylindrical, will have a 60° included angle on one end, and contains a Morse taper on the other end.

17. Select the answer(s) that **best** identifies the result of turning between centers.

    A. Runout is minimized.
    B. Work can be removed and reinstalled with a high level of repeatability.
    C. Dead centers and live centers both contain rotating parts.
    D. All of the above are correct.
    E. Both A and B are correct.

    Answer E is correct. Both answers A and B are results of turning between centers. When turning between centers, runout is minimized and work can be removed and reinstalled with a high level of repeatability.

18. Select the answer that is **not** a characteristic of a universal chuck.

    A. The movement of all the jaws is synchronized.
    B. Each jaw is capable of being moved independently.
    C. The jaws are "self-centering" and move in and out by turning a chuck key.
    D. There are two-, three-, or six-piece jaw configurations.

    Answer B is correct. The jaws do not move independently on a universal chuck, but move simultaneously and are equidistant from the center. Therefore, the universal chuck will "center" the workpiece as the chuck key tightens the jaws.

19. Select the workholding device that is defined incorrectly.

    A. Faceplate—mounted to the headstock spindle and is used for irregular-shaped work
    B. Universal chuck—jaws open and close simultaneously
    C. Collet—accurate, repeatable, and quick to interchange parts
    D. Four-jaw chuck—can be used for eccentric or off-center workholding of triangular-shaped workpieces

    Answer D is correct. A four-jaw chuck can be used for eccentric or off-center work, but could not be used to hold a triangular-shaped workpiece.

20. A machinist must remove 0.025″ from a 1.500″ Ø × 25.00″–long shaft. To safely and accurately complete this task, the setup should include:

    A. a follower rest.
    B. the use of the half-nut lever.
    C. a taper reducer.
    D. a solid mandrel.

    Answer A is correct. A follower rest would provide support for a long workpiece that lacks rigidity.

21. Which workholding device would include the following operating sequence: Open the jaws slightly larger than the diameter of the stock, number each jaw with a marker, lightly clamp the stock while visibly aligning it to center, and use a dial indicator to check/adjust until the desired TIR is achieved?

    A. Faceplate
    B. Universal chuck
    C. Independent chuck
    D. Step collet

    Answer C is correct. This sequence describes the process for centering a part in an independent chuck.

22. The machinist engages the half-nut lever. Which of the following specifications corresponds to this action?

    A. 6-32 UNF-2B
    B. Ø 0.750″ ± 0.005
    C. 1/2″ TPI
    D. 1/2-13 UNC-2A

    Answer D is correct. This specification calls for a thread that could be produced with a single-point cutting tool. This operation would require the half-nut lever to be utilized.

23. The carriage of a lathe is moving 0.003″ per revolution of the spindle. This movement is determined by:

    A. spindle clutch lever.
    B. quick change gear box.
    C. carriage traverse lever.
    D. step pulley.

    Answer B is correct. The knobs and levers connected to the quick change gearbox are set to determine the desired feed rate.

24. The cross slide moves in a direction _____ to the face of the part and at a speed _____ to that of the longitudinal feed.

    A. parallel; one-third to one-half
    B. parallel; equal
    C. perpendicular; one-third to one-half
    D. perpendicular; two times

    Answer A is correct. The cross slide moves parallel to the face of the workpiece, and most lathes at a speed one-third to one-half to the longitudinal feed rate.

25. The size of a lathe is determined by the _____ and the _____.

    A. horsepower; swing
    B. swing; bed length
    C. length; width
    D. maximum allowable tool size; maximum RPM

    Answer B is correct. The swing or biggest possible diameter that can be held in the spindle while clearing the ways, and the length of the bed are the two factors used to specify a lathe's size.

26. Which of the following **best** describes the characteristics of a taper?

    A. Measured in degrees, taper per inch, taper per foot
    B. Continuous change in diameter
    C. Can be self-holding or self-releasing
    D. All of the above

    Answer D is correct. A taper is defined as a continuous change in diameter and is measured in degrees, taper per inch, and taper per foot, and can be self-holding or self-releasing.

27. The RPM for a reaming operation is approximately _____ the RPM of a drilling operation of the same diameter.

    A. one-half
    B. one-fourth
    C. equal to
    D. two times

    Answer A is correct. If a cutting speed chart is not available, one-half the speed of the same-size drilling operation is a good approximation.

28. A facing tool is reinstalled after it is sharpened. The first cut produces a **cone-shaped** protrusion at the center of the piece. What is the probable root cause?

    A. The tool is on center.

    B. The tool is below center.

    C. The tool is above center.

    D. The RPM is set too high.

    Answer C is correct. With an above-center facing tool, a cone-shaped protrusion will be produced by the face of the tool.

29. A raised pattern on the surface of a workpiece is created by displacing the material with two hardened wheels. The basic patterns created by this process are straight and diamond. Identify this process.

    A. Gripping

    B. Pressing

    C. Knurling

    D. Etching

    Answer C is correct. Knurling is the machining process that produces a raised pattern on the surface of the workpiece.

30. Which of the following statements is true about the initial cut on a 3.0″ cast iron shaft?

    A. A generous amount of cutting oil should be applied.

    B. The depth of cut should penetrate below the rough and hard outer scale.

    C. Reduce the cutting speed.

    D. None of the above.

    Answer B is correct. The rough outer scale is tough to machine. Cutting at a depth below this scale will improve machinability and reduce tool wear.

31. The control charts used with SPC to analyze trends in part variation are:

    A. X bar and R charts.

    B. production and cost charts.

    C. capability and first-piece inspection charts.

    D. X graph and Y charts.

    Answer A is correct. X-bar charts track the mean size of a dimension from a sampling of parts. R charts track the amount of variation of a dimension from a sampling of parts. Both are used with SPC.

32. An R chart or range chart shows the variation of each sampling. What is the definition of range in this example?

    A. The average size of a dimension in a sampling

    B. The average time it takes to machine a part

    C. A value obtained by subtracting the smallest dimension from the largest dimension in a sample data set

    D. The maximum distance a measuring tool is capable of measuring

    Answer C is correct. The variation from the smallest value to the largest value is the range.

33. Select the most reliable method described here for repeatable results when turning the length of a shoulder on the lathe that is not equipped with a DRO.

    A. Set the compound rest parallel to the centerline of the lathe.
    B. Set the compound rest perpendicular to the centerline of the lathe.
    C. Support the piece with a tailstock.
    D. Install a micrometer stop to the right of the tailstock.

    Answer A is correct. The micrometer dial on the compound rest will give the machinist repeatable, longitudinal control over the length of the shoulder.

34. The upper left-hand area of the lathe identified by letter B is the _____, which houses the _____ and the _____.

    A. motor; belts; drive screw
    B. gears; clutch; sleeve
    C. cabinet; transmission; collar
    D. headstock; gears; spindle

    Answer D is correct. Letter B is pointing to the headstock, which houses gears that transmit power to the spindle.

35. Letter C identifies the:

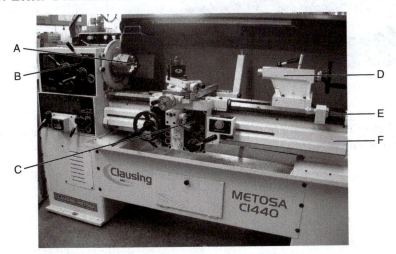

    A. apron.
    B. table.
    C. tool carrier.
    D. compound rest.

    Answer A is correct. Letter C is pointing to the apron, which is attached to the saddle.

36. The carriage slides along the _____, which are identified by letter E.

A. rails
B. tracks

C. guides
D. ways

Answer D is correct. Letter E is pointing to the ways.

37. Letter F represents _____, the casting that runs the length of the lathe.

A. base
B. bed

C. frame
D. column

Answer B is correct. Letter F is pointing to the bed, which is a machined casting that serves as a foundation to the machine.

38. Letter H identifies the _____, which rotates _____.

A. compound rest; 360°   C. compound rest; 180°
B. cross slide; 360°   D. tool post; 180°

Answer A is correct. Letter H is pointing to the compound rest, which is capable of rotating 360° to perform various turning operations.

39. Letter I identifies the:

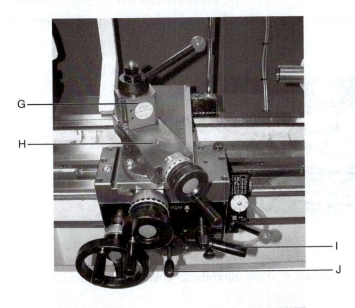

A. half-nut lever.   C. spindle start/stop lever.
B. feed lever.   D. thread retract lever.

Answer A is correct. Letter I is pointing to the half-nut lever, which engages the feed for threading operations.

40. Letter J identifies the:

A. half-nut lever.

B. feed lever.

C. spindle start/stop lever.

D. thread retract lever.

Letter B is correct. Letter J is pointing to the feed lever, which engages the movement of the cross slide or the carriage.

41. Letter K identifies the:

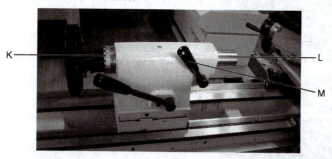

A. measurement dial.

B. dial indicator.

C. fine drilling gage.

D. micrometer collar.

Answer D is correct. Letter K is pointing to the micrometer collar on the tailstock.

42. Letter L identifies the:

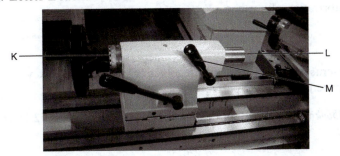

A. ram.

B. tailstock piston.

C. tailstock quill.

D. subspindle.

Answer C is correct. Letter L is pointing to the tailstock quill.

43. Letter M identifies the:

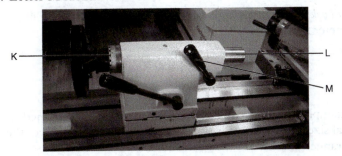

A. feed lever.

B. tailstock quill lock.

C. tailstock retract lever.

D. half-nut lever.

Answer B is correct. Letter M is pointing to the tailstock quill-locking lever.

44. A tailstock that is offset toward the operator will produce a turned workpiece that is:

A. tapered with the end near the tailstock being smaller.

B. undersize the entire length.

C. tapered with the end near the headstock being smaller.

D. oversize the entire length.

Answer A is correct. When turning between centers, a tailstock shifted toward the operator will position the right-hand end of the workpiece closer to the tool than the spindle end of the workpiece. Turning a workpiece setup in this manner will result in a taper, with the right-hand end being smaller than the spindle end.

45. Precise alignment of the centers on a lathe while turning a workpiece can be accomplished by:

A. visually aligning the graduation marks on the base of the tailstock.

B. using a test bar and adjusting a test indicator as necessary.

C. taking a trial cut, measuring each end with a micrometer, and adjusting.

D. bringing the centers together by sliding the tailstock forward, adjusting, and checking visually.

Answer C is correct. Checking the piece for taper after taking a trial cut and then adjusting as necessary is a precise method that can be accomplished while turning the workpiece.

46. Select the acronym that represents the class of fit that is intended to provide similar running capability while providing for lubrication allowance.

    A. FN
    B. RC
    C. LC
    D. LT

    Answer B is correct. RC represents running or sliding clearance fit, for which the key characteristic is free motion between the mating parts.

47. Refer to *The Machinery's Handbook* to select the applicable shaft diameter for an FN2 force fit with a 2.625"-diameter hole:

    A. 2.6225".
    B. 2.6253".
    C. 2.6262".
    D. 2.6275".

    Answer D is correct. In this example the chart is specifying that the shaft should be machined 0.0022" to 0.0029" over the nominal size, so an FN2 fit is obtained. 2.6275" is 0.0025" over the nominal size and is within this range.

48. Refer to *The Machinery's Handbook* to select the applicable bore diameter for an RC4 clearance fit with a 1.300"-diameter shaft.

    A. 1.3010"
    B. 1.3030"
    C. 1.2980"
    D. 1.2990"
    E. 0.0016

    Answer A is correct. In this example the chart is specifying that the bore be machined to 0.0 to 0.0016" over the nominal size so an RC4 fit is obtained. 1.3010" is 0.001" over the nominal size and is within this range.

49. Refer to *The Machinery's Handbook* to determine the appropriate tap drill to produce a tapered 1/4-18 NPT:

    A. 37/64 drill.
    B. 7/32 drill.
    C. 7/16 drill.
    D. 29/64 drill.

    Answer C is correct. Referencing the Tap Drill chart for pipe threads, a 7/16 drill is the correct tap drill to produce this thread.

50. Which of the following serves as a reference plane when used with a height gage to produce precision layouts?

    A. Surface plate
    B. Angle plate
    C. Surface gage
    D. Protractor

    Answer A is correct. A surface plate will provide the reference surface for the height gage to be set from.

Grinding Skills Certification

7.1 PRACTICE THEORY EXAM—GRINDING SKILLS

1. From the abrasive grain sizes listed here, which one is the finest (least coarse)?

   A. 36
   B. 60
   C. 100
   D. 120

2. The condition in which deposits of workpiece material become embedded in the surface of a grinding wheel, leading to poor cutting conditions, describes:

   A. pinning.
   B. dressing.
   C. bonding.
   D. loading.

3. To prepare an installed surface grinding wheel for rough grinding, the machinist should:

   A. adjust the down-feed depth 0.003″.
   B. dress with a star dresser.
   C. feed the dresser at a fast rate.
   D. cross-feed the dresser at a slow rate.
   E. A and C are correct.

4. A part print specifies a part's height to be $0.7250 + 0.0003″ - 0.0000″$ and a quantity of 50 parts. Which of the following is an accurate and efficient method to inspect this dimension during the machining process?

   A. Depth micrometer and surface plate
   B. 0–1″ micrometer
   C. Height gage and surface plate
   D. Dial indicator mounted to a surface gage with a surface plate, calibrated with gage blocks

5. Which of the following statements are **true** about aluminum oxide grinding wheels?

   A. The grit type designated by the letter "A"
   B. A common abrasive type for grinding steel
   C. Can be pink, white, or brown
   D. All of the above
   E. None of the above

6. Nonferrous materials can be held on a magnetic chuck by the following methods, **except**:

   A. with a grinding vise and parallels.
   B. directly to the chuck's surface.
   C. with flexible magna-vise clamps compressed against the sides/ends of the workpiece.
   D. clamped to an angle plate.

7. Prior to starting a pedestal grinder, it is recommended the:

   A. tool rest and spark arrestor are checked and adjusted.

   B. operator puts work gloves on.

   C. operator stands to the side of the grinder.

   D. All of the above.

   E. A and C are correct.

8. A machinist needs to sharpen a tungsten carbide cutting tool. Which of the following grinding wheel choices could be used?

   A. Aluminum oxide

   B. Black—silicon carbide

   C. Green—silicon carbide

   D. Diamond

   E. C and D

9. Letter A identifies the _____ on a pedestal grinder, which should be positioned no more than _____ from the face of the grinding wheel.

   A. tool rest; 1/8″

   B. compound rest; 3/16″

   C. tool holder; 5/16″

   D. wheel guard; 3/8″

10. After installing a new grinding wheel on a surface grinder, the machinist determines the profile of the wheel and the axis of rotation are running eccentric to each other. The process of correcting this condition is called:

    A. indicating.

    B. balancing.

    C. centering.

    D. truing.

11. A machinist is setting up a surface grinder that has a spindle speed of 3450 RPM and a wheel capacity of 8″ × 3/4″. Which wheel(s) listed here is safe and capable of running on this machine?

    A. 8″ Ø × 1/2″ thickness, maximum RPM 3105

    B. 8″ Ø × 1/2″ thickness, maximum RPM 3600

    C. 8″ Ø × 1″ thickness, maximum RPM 3600

    D. 7″ Ø × 3/4″ thickness, maximum RPM 3140

12. After grinding surfaces A and B of the workpiece pictured here and checking the flatness with a dial indicator, the workpiece is 0.0015″ thicker at one end. What could have caused this to occur?

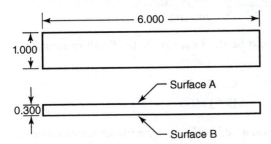

A. A burr or piece of debris is present in between the part and the chuck.

B. The chuck's surface is worn unevenly.

C. The wheel is too hard.

D. All of the above are possible causes.

13. Which of the following answers **best** describes the purpose of the blotters on a grinding wheel?

A. Ensure uniform distribution of the flange pressure

B. Prevent slipping between the flange and the wheel

C. Provide important information about safety and wheel characteristics

D. All of the above

E. A and B

14. A grinding wheel with a _____ bond type is used for most general grinding operations.

A. vitrified

B. hard

C. resinoid

D. silicate

15. The spacing between the individual grains on a grinding wheel refers to _____ and is identified by a number value from 1 to 16.

A. structure

B. bond

C. pitch

D. grade

16. A reasonable depth of cut for a finishing operation on a surface grinder is:

A. 0.0001″–0.0005″

B. 0.0005″–0.001″

C. 0.001″–0.0015″

D. 0.0015″–0.002″

17. After removing 0.005″ from the height of a workpiece on a surface grinder, burn marks start to appear on the surface. This is an indication of which of the following?

A. The face of the wheel may be glazed and needs to be redressed.

B. The cross-feed step-over amount should be increased.

C. The magnetic chuck is wearing unevenly.

D. Both A and B are correct.

18. Which of the following statements **best** describes the grade of a grinding wheel?

A. Grinding wheel are designated by letters A (softest) through Z (hardest).

B. Softer-grade wheels are best for grinding harder materials.

C. Harder wheels are more wear resistant.

D. All of the above are correct.

E. A and B are correct.

19. A technique for preventing a workpiece from tipping or sliding on the magnetic chuck by placing blocks and/or parallels around the workpiece is called:

    A. squaring.

    B. blocking.

    C. clamping.

    D. pinning.

20. Which of the grinding wheels listed here would be the **best** choice for finish grinding carbide?

    A. D126

    B. A100

    C. C120

    D. D400

21. Which of the following should **not** be done on a pedestal grinder with an aluminum oxide wheel?

    A. Using a grinding wheel with an RPM rating that is less than the speed of the grinder

    B. Wearing gloves to hold the workpiece

    C. Grinding aluminum or brass

    D. All of the above

22. A machinist is preparing to set up an irregularly shaped workpiece on a surface grinder. Which workholding device will support the irregular shape, while extending the magnetic field of the magnetic chuck?

    A. Sine plate

    B. Grinding vise

    C. Magnetic parallels

    D. Angle plate

23. The correct device for dressing a diamond grinding wheel is a:

    A. brake truing device.

    B. aluminum oxide dressing stick.

    C. single-point diamond dresser.

    D. cluster-point dresser.

24. Which of the following statements is **untrue** about the use of coolant during surface grinding operations?

    A. Provides greater lubricity between the wheel and the part

    B. Flushes metal particles and released wheel grains

    C. Lowers the temperature of the workpiece

    D. Removes glazing from the wheel's surface

    E. Facilitates better surface finishes

25. A surface grinder wheel that is out of round and/or out of balance could result in:

    A. burn marks.

    B. a wavy surface.

    C. scratches on the surface.

    D. None of the above.

26. What does the 38 identify in the grinding wheel specification 38 A 80 I 8 V?

    A. The grain size

    B. The grade

    C. A manufacturer's prefix

    D. The hub diameter

27. The J on the grinding wheel specification in the following image designates which grinding wheel characteristic?

A. Abrasive type

B. Bond type

C. Grade/hardness

D. Structure

28. The process of correcting irregularities in the shape of a surface grinding wheel refers to _____ and is completed with a dressing stick or wheel dresser.

A. aligning

B. balancing

C. truing

D. ringing

29. A surface grinding technique that involves taking multiple passes over the workpiece until sparks are no longer produced is referred to as:

A. plunge grinding.

B. spring pass.

C. sparking out.

D. relief cutting.

30. Which of the following grinding wheels would be the **best** choice for grinding hardened tool steel?

A. Silicon carbide

B. Cubic boron nitride

C. Ceramic aluminum oxide

D. None of the above

31. Which of the following statement(s) below identifies the advantages of a magnetic chuck, in comparison to most mechanical workholding devices?

A. Slow setup and tear down

B. Consistent clamping force over the surface of the workpiece

C. Full support under the workpiece

D. Limited access to the work surface

E. B and C

32. A machine tool with a horizontal spindle that rotates a grinding wheel and removes material as the workpiece is reciprocated under the wheel is a:

A. cylindrical grinder.

B. surface grinder.

C. tool and cutter grinder.

D. die grinder.

33. Which material listed here would a pedestal grinder equipped with a silicon carbide wheel be used to grind?

A. Cast iron

B. Carbide

C. Stainless steel

D. Nonferrous metals

E. All of the above

34. To ensure even wear of the point on a diamond dresser, the machinist should regularly:

A. demagnetize the dresser.

B. polish the diamond.

C. rotate the diamond in the holder.

D. All of the above

35. A light cut (0.0001″–0.0005″), an open wheel structure, and flood coolant utilized during a surface grinding operation will increase the opportunity for a:

A. flat workpiece.

B. true and concentric wheel.

C. smooth surface finish.

D. deeper depth of cut.

36. _____ describes the material removal process performed by the edges of the individual grains on the surface of a grinding wheel.

A. Lapping

B. Honing

C. Milling

D. Cutting

37. Which of the following abrasive types would be applicable for grinding brass or other soft nonferrous metals?

A. Aluminum oxide

B. Diamond

C. Silicon carbide

D. None of the above

38. To avoid uneven wear of the magnetic chuck, the machinist should:

A. hold the workpiece in a grinding vise.

B. secure the workpiece on magnetic parallels.

C. use a different area of the magnetic chuck with each setup.

D. limit the down-feed to 0.0002″.

39. Which component on the grinder in the following image moves the workpiece longitudinally?

A. A

B. B

C. C

D. D

40. Which component on the grinder in the following image moves the wheel up and down?

A. A

B. B

C. C

D. D

41. Letter E in the following image identifies the:

A. wheel guard.

B. knee.

C. magnetic chuck.

D. saddle.

42. Letter D in the following image identifies the:

    A. longitudinal handwheel.
    B. cross-feed handwheel.
    C. elevating handwheel.
    D. None of the above.

43. Ring-testing a grinding wheel will identify which of the following?

    A. Wheel balance
    B. A cracked wheel
    C. Concentricity of the bore
    D. Wheel hardness

44. The hardness of a grinding wheel is determined by the:

    A. grade.
    B. bond.
    C. structure.
    D. grain size.

45. Aluminum oxide is a _____ and is applicable for grinding common types of steel.

    A. bonding material
    B. grain type
    C. superabrasive
    D. None of the above

46. Which statement is **not** true regarding diamond abrasive wheels?

    A. Dressed with a single-point diamond dresser
    B. Hardest natural abrasive
    C. Used to grind carbide
    D. Specified by the letter D

47. When dressing a wheel on a surface grinder, to obtain a closed wheel structure for finish grinding, the machinist should:

    A. cross-feed the dresser slowly.

    B. feed the cross feed at a fast rate.

    C. lower the head 0.0002"–0.0005" per pass.

    D. All of the above.

    E. A and C

48. What type of grinding wheel listed here has a bonding material with some flexibility and is capable of withstanding pressure? These characteristics make this type of bond applicable for thin wheel grinding operations.

    A. Vitrified

    B. Shellac

    C. Rubber

    D. Resin

49. The symbol 30$\sqrt{}$ indicates that _____ and is _____ in comparison to a 125$\sqrt{}$.

    A. a surface finish of 30 microinches or better is required; better

    B. the carbon percent is 30%; softer

    C. a surface finish of 30 microinches or better is required; worse

    D. the carbon percent is 30%; harder

50. Running the coolant on a stopped grinding wheel could result in:

    A. the coolant washing the color out of the wheel.

    B. the pores of the wheel absorbing the coolant and the wheel running out of balance when started.

    C. eroding of the bonding material.

    D. All of the above.

7.2 ANSWER KEY

1. D
2. D
3. E
4. D
5. D
6. B
7. E
8. E
9. A
10. D
11. B
12. D
13. D
14. A
15. A
16. A
17. A
18. D
19. B
20. D
21. D
22. C
23. B
24. D
25. B

26. C
27. C
28. C
29. C
30. B
31. E
32. B
33. E
34. C
35. C
36. D
37. C
38. C
39. C
40. A
41. C
42. D
43. B
44. A
45. B
46. A
47. E
48. C
49. A
50. B

7.3 PRACTICE THEORY EXAM EXPLANATIONS

1. From the abrasive grain sizes listed here, which one is the finest (least coarse)?

    A. 36   C. 100

    B. 60   D. 120

    Answer D is correct. When specifying abrasive grain sizes, the larger the grain number the smaller in size the grain particle is. In this example 120 is the smallest grain size, and 36 is the largest.

2. The condition in which deposits of workpiece material become embedded in the surface of a grinding wheel, leading to poor cutting conditions, describes:

    A. pinning.   C. bonding.

    B. dressing.   D. loading.

    Answer D is correct. The metal particles or chips produced during surface grinding operations can become embedded in the voids on the face of a grinding wheel. Loading reduces the contact area between the individual grains and the workpiece, resulting in greater friction and heat buildup.

3. To prepare an installed surface grinding wheel for rough grinding, the machinist should:

    A. adjust the down-feed depth 0.003″.   D. cross-feed the dresser at a slow rate.

    B. dress with a star dresser.   E. A and C are correct.

    C. feed the dresser at a fast rate.

    Answer E is correct. A depth of approximately 0.003″ and a fast feed rate will expose fresh abrasive grains and "open" the face of the wheel. This will provide a more aggressive cutting surface for rough grinding.

4. A part print specifies a part's height to be 0.7250 + 0.0003″ − 0.0000″ and a quantity of 50 parts. Which of the following is an accurate and efficient method to inspect this dimension during the machining process?

    A. Depth micrometer and surface plate   D. Dial indicator mounted to a surface

    B. 0–1″ micrometer   gage with a surface plate, calibrated

    C. Height gage and surface plate   with gage blocks

    Answer D is correct. "Zeroing" a dial indicator mounted on a surface gage to the height of the 0.7520″ gage blocks will provide a reference. The manufactured parts can then easily be checked by sliding the parts under the indicator. As long as a part's variation to the "zeroed" indicator does not exceed the tolerance, it is a good part.

5. Which of the following statements are **true** about aluminum oxide grinding wheels?

    A. The grit type designated by the   C. Can be pink, white, or brown
       letter "A"   D. All of the above

    B. A common abrasive type for grinding   E. None of the above
       steel

    Answer D is correct. Aluminum oxide grinding wheels are commonly used for grinding steel and are designated by the letter A in the second characteristic of the grinding wheel marking system on a grinding wheel blotter. Additives such as chromium or titanium will change the color to pink or brown. A pure aluminum oxide wheel will be white.

6. Nonferrous materials can be held on a magnetic chuck by the following methods, **except**:

    A. with a grinding vise and parallels.

    B. directly to the chuck's surface.

    C. with flexible magna-vise clamps compressed against the sides/ends of the workpiece.

    D. clamped to an angle plate.

    Answer B is correct. Nonmagnetic materials cannot be secured solely to the magnetic chuck without additional mechanical workholding devices.

7. Prior to starting a pedestal grinder, it is recommended the:

    A. tool rest and spark arrestor are checked and adjusted.

    B. operator puts work gloves on.

    C. operator stands to the side of the grinder.

    D. All of the above.

    E. A and C are correct.

    Answer E is correct. To operate a pedestal grinder safely, the tool rest and spark arrestor should be set at an appropriate distance from the face of the wheel. The operator should stand to the side when starting the grinder in an effort to stay out of harm's way, should any problems occur with the wheel.

8. A machinist needs to sharpen a tungsten carbide cutting tool. Which of the following grinding wheel choices could be used?

    A. Aluminum oxide

    B. Black—silicon carbide

    C. Green—silicon carbide

    D. Diamond

    E. C and D

    Answer E is correct. Silicon carbide and diamond grinding wheels contain abrasives with a hardness capable of effectively cutting hard materials like carbide.

9. Letter A identifies the _____ on a pedestal grinder, which should be positioned no more than _____ from the face of the grinding wheel.

    A. tool rest; 1/8″

    B. compound rest; 3/16″

    C. tool holder; 5/16″

    D. wheel guard; 3/8″

    Answer A is correct. The tool rest provides a stable platform to place the workpiece when offhand grinding. To minimize the opportunity for serious injury, the tool rest should ideally be adjusted to 1/16″ and no more than 1/8″ from the face of the wheel.

10. After installing a new grinding wheel on a surface grinder, the machinist determines the profile of the wheel and the axis of rotation are running eccentric to each other. The process of correcting this condition is called:

    A. indicating.
    B. balancing.
    C. centering.
    D. truing.

    Answer D is correct. Truing is the process of removing bond material and abrasive grains from the surface of the grinding wheel to correct the wheel's irregular shape.

11. A machinist is setting up a surface grinder that has a spindle speed of 3450 RPM and a wheel capacity of 8″ × 3/4″. Which wheel(s) listed here is safe and capable of running on this machine?

    A. 8″ Ø × 1/2″ thickness, maximum RPM 3105
    B. 8″ Ø × 1/2″ thickness, maximum RPM 3600
    C. 8″ Ø × 1″ thickness, maximum RPM 3600
    D. 7″ Ø × 3/4″ thickness, maximum RPM 3140

    Answer B is correct. The size specification of the wheel indicated in answer B is within the capacity of this grinder and the RPM rating is higher than the speed of the spindle.

12. After grinding surfaces A and B of the workpiece pictured here and checking the flatness with a dial indicator, the workpiece is 0.0015″ thicker at one end. What could have caused this to occur?

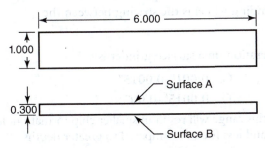

    A. A burr or piece of debris is present in between the part and the chuck.
    B. The chuck's surface is worn unevenly.
    C. The wheel is too hard.
    D. All of the above are possible causes.

    Answer D is correct. Several factors could contribute to a surface-ground workpiece not meeting flatness requirements. A burr or piece of debris underneath the workpiece will elevate that end or section. This elevated state will result in more material being removed from that portion of the workpiece than the portion secured directly to the magnetic chuck. An uneven surface on the magnetic chuck could cause the workpiece to pull down to the chuck in a distorted manner. This distortion will result in variations on the finished surfaces of the workpiece. Depending on the workpiece material, a wheel that is too hard will not release grains, which will in turn become dull. This will result in an increase in friction and heat, which can warp the workpiece.

13. Which of the following answers **best** describes the purpose of the blotters on a grinding wheel?

    A. Ensure uniform distribution of the flange pressure

    B. Prevent slipping between the flange and the wheel

    C. Provide important information about safety and wheel characteristics

    D. All of the above

    E. A and B

    Answer D is correct. The blotter serves several purposes, which include providing a cushion and a greater coefficient of friction between the flange and the uneven surface of the wheel, and it shows the printed specifications of the wheel.

14. A grinding wheel with a _____ bond type is used for most general grinding operations.

    A. vitrified

    B. hard

    C. resinoid

    D. silicate

    Answer A is correct. A vitrified bonding material is a glass-type resin that holds the grains together. This type of bond is the most common for precision grinding operations.

15. The spacing between the individual grains on a grinding wheel refers to _____ and is identified by a number value from 1 to 16.

    A. structure

    B. bond

    C. pitch

    D. grade

    Answer A is correct. The structure of a grinding wheel is the spacing between the individual grains.

16. A reasonable depth of cut for a finishing operation on a surface grinder is:

    A. 0.0001″–0.0005″

    B. 0.0005″–0.001″

    C. 0.001″–0.0015″

    D. 0.0015″–0.002″

    Answer A is correct. A depth of cut within this range will result in smaller chip formation, less grinding forces applied to the part's surface, and less friction compared to greater depths.

17. After removing 0.005″ from the height of a workpiece on a surface grinder, burn marks start to appear on the surface. This is an indication of which of the following?

    A. The face of the wheel may be glazed and needs to be redressed.

    B. The cross-feed step-over amount should be increased.

    C. The magnetic chuck is wearing unevenly.

    D. Both A and B are correct.

    Answer A is correct. The condition in which the voids between the abrasive grains on the surface of a grinding wheel become clogged with workpiece material describes glazing. Grinding with a glazed wheel will minimize the exposure and cutting edges of the abrasive grains, which will in turn lead to excessive heat and burning.

18. Which of the following statements **best** describes the grade of a grinding wheel?

    A. Grinding wheel are designated by letters A (softest) through Z (hardest).

    B. Softer-grade wheels are best for grinding harder materials.

    C. Harder wheels are more wear resistant.

    D. All of the above are correct.

    E. A and B are correct.

    Answer D is correct. The grade of a grinding wheel is a measure of the strength of its bond, specified by letters A through Z. A softer wheel will release old, dull grains at an appropriate rate when grinding harder materials. A harder wheel is most applicable when removing large amounts of material because it is more wear resistant.

19. A technique for preventing a workpiece from tipping or sliding on the magnetic chuck by placing blocks and/or parallels around the workpiece is called:

    A. squaring.

    B. blocking.

    C. clamping.

    D. pinning.

    Answer B is correct. When a part is taller than it is wide, it may lack stability when secured to the magnetic chuck. This could lead to its toppling during grinding operations. Placing blocks around the workpiece will provide support against the force applied by the grinding wheel.

20. Which of the grinding wheels listed here would be the **best** choice for finish grinding carbide?

    A. D126

    B. A100

    C. C120

    D. D400

    Answer D is correct. A diamond wheel is appropriate for grinding carbide and is designated by the letter D on the grinding wheel marking system. For finish grinding carbide, a grain size of 400 is a good choice to achieve a nice surface finish.

21. Which of the following should **not** be done on a pedestal grinder with an aluminum oxide wheel?

    A. Using a grinding wheel with an RPM rating that is less than the speed of the grinder

    B. Wearing gloves to hold the workpiece

    C. Grinding aluminum or brass

    D. All of the above

    Answer D is correct. When working on a pedestal grinder, the machinist must verify that the RPM rating for the grinding wheel exceeds the RPM of the grinder. Wearing gloves while grinding could lead to serious injury if the glove were to catch on the wheel and pull the machinist's hand into it. Aluminum oxide grinding wheels are not well suited for grinding soft materials such as aluminum or brass.

22. A machinist is preparing to set up an irregularly shaped workpiece on a surface grinder. Which workholding device will support the irregular shape, while extending the magnetic field of the magnetic chuck?

    A. Sine plate

    B. Grinding vise

    C. Magnetic parallels

    D. Angle plate

    Answer C is correct. Magnetic parallels are blocks designed to transfer or carry the magnetism generated by the magnetic chuck. They can be positioned under a workpiece to avoid having deviations or projections on the underside (side facing the magnetic chuck) affect the setup.

23. The correct device for dressing a diamond grinding wheel is a:

    A. brake truing device

    B. aluminum oxide dressing stick

    C. single-point diamond dresser

    D. cluster-point dresser

    Answer B is correct. To expose fresh grains on the face of a diamond grinding wheel, an aluminum oxide dressing stick should be moved back and forth across this surface.

24. Which of the following statements is **untrue** about the use of coolant during surface grinding operations?

    A. Provides greater lubricity between the wheel and the part

    B. Flushes metal particles and released wheel grains

    C. Lowers the temperature of the workpiece

    D. Removes glazing from the wheel's surface

    E. Facilitates better surface finishes

    Answer D is correct. Coolant will not remove workpiece material and particles that have become embedded in the voids on a wheel's surface. To expose fresh grains and remove the glazing, the wheel should be dressed.

25. A surface grinder wheel that is out of round and/or out of balance could result in:

    A. burn marks.

    B. a wavy surface.

    C. scratches on the surface.

    D. None of the above.

    Answer B is correct. An out-of-round grinding wheel will have significant radial runout, which will create vibration in the wheel. This will lead to intermittent cutting action as the high spots on the wheel make contact with the workpiece. The end result will be a wavy, uneven surface finish.

26. What does the 38 identify in the grinding wheel specification 38 A 80 I 8 V?

    A. The grain size

    B. The grade

    C. A manufacturer's prefix

    D. The hub diameter

    Answer C is correct. The first characteristic on the grinding wheel marking system is a manufacturer's code for the abrasive.

27. The J on the grinding wheel specification in the following image designates which grinding wheel characteristic?

    A. Abrasive type

    B. Bond type

    C. Grade/hardness

    D. Structure

    Answer C is correct. The letter J, which is the fourth characteristic on the grinding wheel marking system, identifies the grade of the wheel. Grade is specified by letters A–Z, with A being the softest and Z being the hardest.

28. The process of correcting irregularities in the shape of a surface grinding wheel refers to _____ and is completed with a dressing stick or wheel dresser.

    A. aligning
    B. balancing
    C. truing
    D. ringing

    Answer C is correct. Truing is the process of removing high spots and irregularities from the surface grinding wheel to achieve concentricity between the periphery of the wheel and the axis of rotation. Truing can be accomplished simultaneously through the wheel dressing process.

29. A surface grinding technique that involves taking multiple passes over the workpiece until sparks are no longer produced is referred to as:

    A. plunge grinding.
    B. spring pass.
    C. sparking out.
    D. relief cutting.

    Answer C is correct. "Sparking out" or intentionally traversing the grinding wheel over the workpiece without increasing the cutting depth is a technique used to achieve a good surface finish and ensure all the surface's material is removed at the set height.

30. Which of the following grinding wheels would be the **best** choice for grinding hardened tool steel?

    A. Silicon carbide
    B. Cubic boron nitride
    C. Ceramic aluminum oxide
    D. None of the above

    Answer B is correct. Cubic boron nitride (CBN) grinding wheels are hard, have high strength, and are wear resistant. These characteristics make CBN wheels effective for grinding hard materials.

31. Which of the following statement(s) below identifies the advantages of a magnetic chuck, in comparison to most mechanical workholding devices?

    A. Slow setup and tear down
    B. Consistent clamping force over the surface of the workpiece
    C. Full support under the workpiece
    D. Limited access to the work surface
    E. B and C

    Answer E is correct. A magnetic chuck is a versatile workholding device that provides several advantages over mechanical workholding devices. These advantages include quick setup and tear-down time, clamping force and support applied over the majority of the workpiece, and unobstructed access to the workpiece surface.

32. A machine tool with a horizontal spindle that rotates a grinding wheel and removes material as the workpiece is reciprocated under the wheel is a:

    A. cylindrical grinder.
    B. surface grinder.
    C. tool and cutter grinder.
    D. die grinder.

    Answer B is correct. A surface grinder typically has a horizontal spindle on which a grinding wheel is mounted. The secured workpiece is traversed by a reciprocating table under the rotating grinding wheel.

33. Which material listed here would a pedestal grinder equipped with a silicon carbide wheel be used to grind?

    A. Cast iron
    B. Carbide
    C. Stainless steel
    D. Nonferrous metals
    E. All of the above

    Answer E is correct. The sharpness, durability, and hardness of silicon carbide grinding wheels make them applicable for grinding cast iron, carbide, stainless steel, and nonferrous materials.

34. To ensure even wear of the point on a diamond dresser, the machinist should regularly:

    A. demagnetize the dresser.
    B. polish the diamond.
    C. rotate the diamond in the holder.
    D. All of the above.

    Answer C is correct. Repeatedly using the same side of a single-point diamond dresser will lead to uneven wear on the diamond. Regularly rotating the diamond in the holder will facilitate even wear.

35. A light cut (0.0001″–0.0005″), an open wheel structure, and flood coolant utilized during a surface grinding operation will increase the opportunity for a:

    A. flat workpiece.
    B. true and concentric wheel.
    C. smooth surface finish.
    D. deeper depth of cut.

    Answer C is correct. A light depth of cut, an open wheel structure to reduce the surface contact between the wheel and the workpiece, and flooding the piece with coolant to wash away wheel grains and workpiece material are grinding techniques that will help to obtain a smooth surface finish.

36. _____ describes the material removal process performed by the edges of the individual grains on the surface of a grinding wheel.

    A. Lapping
    B. Honing
    C. Milling
    D. Cutting

    Answer D is correct. The individual grains on the surface of a grinding wheel cut the metal as they pass by the surface of the work.

37. Which of the following abrasive types would be applicable for grinding brass or other soft nonferrous metals?

    A. Aluminum oxide
    B. Diamond
    C. Silicon carbide
    D. None of the above

    Answer C is correct. The sharpness, durability, and hardness of silicon carbide grinding wheels make them applicable for grinding soft nonferrous materials.

38. To avoid uneven wear of the magnetic chuck, the machinist should:

    A. hold the workpiece in a grinding vise.
    B. secure the workpiece on magnetic parallels.
    C. use a different area of the magnetic chuck with each setup.
    D. limit the down-feed to 0.0002″.

    Answer C is correct. Repeatedly using the same area of a magnetic chuck could lead to accelerated wear at that particular location on the chuck. Using different areas of the chuck will facilitate even wear of the chuck's surface.

39. Which component on the grinder in the following image moves the workpiece
    longitudinally?

A. A   C. C

B. B   D. D

Answer C is correct. Letter C is pointing to the longitudinal feed handwheel, which when
rotated moves the table back and forth.

40. Which component on the grinder in the following image moves the wheel up and down?

A. A   C. C

B. B   D. D

Answer A is correct. Letter A is pointing to the elevating handwheel, which when rotated
raises and lowers the head of the grinder.

41. Letter E in the following image identifies the:

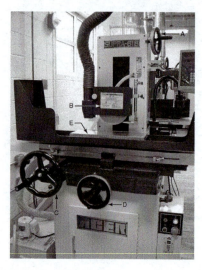

A. wheel guard.   C. magnetic chuck.

B. knee.   D. saddle.

Answer C is correct. Letter E is pointing to the magnetic chuck, which when activated secures ferromagnetic workpieces or workholding devices.

42. Letter D in the following image identifies the:

A. longitudinal handwheel.   C. elevating handwheel.

B. cross-feed handwheel.   D. None of the above.

Answer B is correct. Letter D is pointing to the cross-feed handwheel, which when rotated moves the table in and out.

43. Ring-testing a grinding wheel will identify which of the following?

    A. Wheel balance   C. Concentricity of the bore

    B. A cracked wheel   D. Wheel hardness

    Answer B is correct. A ring test or suspending the wheel by the center hole and tapping it with a blunt object such as the plastic handle on a screwdriver will produce a sound. If the sound is a dull thud, then it is likely the wheel has an internal crack and should be discarded. If the wheel produces a ringing sound, it does not have internal cracks and may be used.

44. The hardness of a grinding wheel is determined by the:

    A. grade.   C. structure.

    B. bond.   D. grain size.

    Answer A is correct. The grade of a grinding wheel is a measure of the strength of its bond, specified by letters A through Z. A is the softest grade and Z is the hardest.

45. Aluminum oxide is a _____ and is applicable for grinding common types of steel.

    A. bonding material   C. superabrasive

    B. grain type   D. None of the above

    Answer B is correct. Aluminum oxide is a common grain type for grinding wheels and is most applicable for grinding steel.

46. Which statement is **not** true regarding diamond abrasive wheels?

    A. Dressed with a single-point diamond   C. Used to grind carbide
       dresser   D. Specified by the letter D

    B. Hardest natural abrasive

    Answer A is correct. A single-point diamond dresser would be the correct tool for dressing an aluminum oxide or silicon carbide grinding wheel. Diamond wheels can be dressed with an aluminum oxide dressing stick.

47. When dressing a wheel on a surface grinder, to obtain a closed wheel structure for finish grinding, the machinist should:

    A. cross-feed the dresser slowly.   D. All of the above

    B. feed the cross feed at a fast rate.   E. A and C

    C. lower the head 0.0002″–0.0005″
       per pass.

    Answer E is correct. A slow cross feed and a shallow dressing depth will ensure that protruding fragments on the face of the wheel will be removed during the wheel dressing process.

48. What type of grinding wheel listed here has a bonding material with some flexibility and is capable of withstanding pressure? These characteristics make this type of bond applicable for thin wheel grinding operations.

    A. Vitrified   C. Rubber

    B. Shellac   D. Resin

    Answer C is correct. A rubber bond type has elasticity and is shock resistant. These characteristics allow the wheel to maintain its integrity when ground to thinner widths.

49. The symbol 30√ indicates that _____ and is _____ in comparison to a 125√.

    A. a surface finish of 30 microinches or better is required; better

    B. the carbon percent is 30%; softer

    C. a surface finish of 30 microinches or better is required; worse

    D. the carbon percent is 30%; harder

    Answer A is correct. Surface finish is measured in microinches, with measurements ranging from 0.5 to 1000. The smaller the value of a surface finish measurement the smoother the surface will be.

50. Running the coolant on a stopped grinding wheel could result in:

    A. the coolant washing the color out of the wheel.

    B. the pores of the wheel absorbing the coolant and the wheel running out of balance when started.

    C. eroding of the bonding material.

    D. All of the above.

    Answer B is correct. A stationary grinding wheel will not deflect the coolant as it would if it were rotating. The coolant could permeate the pores of the stopped wheel, resulting in an unbalanced wheel.

CNC Milling Certifications

8.1 PRACTICE THEORY EXAM—CNC MILLING OPERATIONS/CNC MILLING PROGRAMMING, SETUP, AND OPERATIONS

Questions 1–40 are designed to prepare the student for the CNC operator exam and a portion of the CNC programming, setup, and operations exam. Questions 41–70 are designed to prepare the student for the additional concepts assessed by the CNC programming, setup, and operations exam.

1. Which code activates the workpiece coordinate system?

   A. G43

   B. G41

   C. G17

   D. G54

2. Which code directs the program to end and return to start?

   A. M02

   B. M99

   C. M30

   D. G80

3. Which of the following answers **best** identifies a countersink tool?

   A. A tool used to produce a flat bearing surface on a rough, uneven surface.

   B. A tool used to create an angled bevel to remove burrs, aid the start of taps, and the entry of dowel pins.

   C. A tool used to cut a larger-diameter hole to a specific depth to allow a bolt head or nut to sit flush or below the part's surface.

   D. A tool used to cut a tapered opening at the beginning of an existing hole to a specific depth to allow a flathead screw to sit flush or below the part's surface.

   E. B and D are correct.

4. Typically done as part of the start-up procedure, which is the process of positioning the CNC machining center's axes at a fixed reference position to allow the machine control unit (MCU) to accurately establish/track the table's position?

   A. Jogging

   B. Calibrating

   C. Homing

   D. Positioning

5. To utilize an M01 command, which mode function would have to be selected?

   A. Block delete

   B. Select program

   C. Optional stop

   D. Single block

6. After roughing an internal pocket, a program stop is utilized in the program so the metal chips can be cleared. What will restart the machine from this point in the program?

    A. Power up/restart   C. Cycle start

    B. Power on   D. Reset

7. Which of the following describes a material safety data sheet (MSDS)?

    A. MSDS is used to comply with OSHA's hazard communication standard.
    C. It provides information about safe handling of chemicals and materials.

    B. It identifies the chemicals in a workplace and their recommended uses.
    D. All of the above describe an MSDS.

8. Select the code to fill in the blank. To change tools on a vertical machining center from the tool currently in the spindle to tool number 8 using MDI, the machinist would input _____ T8.

    A. G96   C. M08

    B. M06   D. M02

9. A part print calls for a workpiece to have a finished length of 2.025″. Leaving 0.100″ on the length for machining operations, what is the length of stock that must be cut?

    A. 2 1/16″   C. 2 1/8″

    B. 2 3/32″   D. 2 1/4″

10. A machinist has just set up a VMC and ran a setup piece for a program that has been proven and run in the past. The print in the following image shows the desired hole pattern and the actual hole pattern, which was machined on the setup piece. What is the likely cause of this situation?

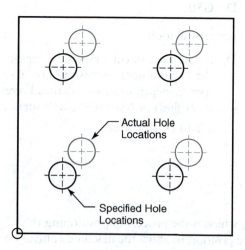

    A. The program coordinates are incorrect.
    C. The tool offset is incorrect.

    B. The work coordinates or fixture offset is incorrect.
    D. There is an M30 in the program.

11. After milling the pocket in the following image, the depth measurement is 0.747″. Which of the following answers will **correctly** adjust this depth when this tool in the program is run again?

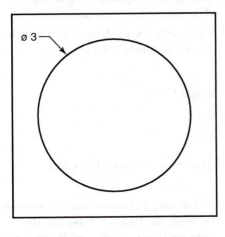 

0.75

A. Change the *Z* value for this tool in the program.

B. Adjust the *Z* wear offset +.003″.

C. Place shims under the workpiece.

D. Adjust the *Z* wear offset −.003″.

E. All of the above

12. You are running a setup piece to produce the part in the following image. The diameter of the pocket finishes at 2.992″. What change to the wear offset will be necessary to machine this pocket to the nominal dimension?

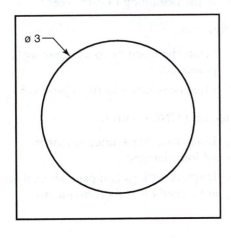 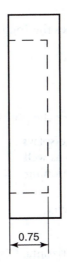

0.75

A. +.005″

B. +.004″

C. −.006″

D. −.004″

13. Which statement **best** describes a datum?

    A. A rectangular box with a series of compartments for symbols and dimensions that describe a part feature

    B. A numerical value positioned within a box used to describe a theoretical dimension

    C. Shown by a capital letter in a box and signifies a plane from which dimensions are referenced

    D. Compares the relationship of two or more cylindrical features

14. Which code identifies linear interpolation (feed in a straight line)?

    A. G00

    B. G01

    C. M01

    D. G20

15. In which of the following scenarios is the individual following recommended safety guidelines?

    A. A machine tool service technician is repairing the machine you operate and is only wearing prescription glasses with no side shields.

    B. The plant manager is watching from the aisle of the shop and is not wearing eye protection.

    C. A coworker puts approved safety glasses on while the tool is cutting and then removes them to measure the workpiece.

    D. You wear safety glasses every time you enter the shop floor and remove them only when you have exited the manufacturing environment.

16. To maintain proper coolant levels and concentrations and to ensure a machine tool has sufficient lubrication, a machinist should check the machine's fluids:

    A. at the end of each week.

    B. every shift.

    C. every other week.

    D. at the beginning of each week.

17. Select the answer that **best** describes the Jog mode function.

    A. Allows programs to be modified

    B. Allows manual control of axes movement

    C. Sends the machine to its reference position

    D. Must be selected to run a program

18. Select the answer that **best** describes the position page on a CNC control.

    A. A digital reference screen that displays work and machine coordinates as well as the distance to go for a positioning command

    B. A screen displaying tool geometry, wear, and work offsets

    C. Lists machine parameters that are seldom changed

    D. Displays settings and parameters that are turned on and off frequently

19. If you are standing in front of a horizontal CNC milling machine, which axis moves in and out (front to back)?

    A. x

    B. y

    C. z

    D. C

20. G codes are referred to as:

    A. miscellaneous commands.

    B. canned cycles.

    C. position codes.

    D. preparatory commands.

21. A machinist is running a setup piece for a part program that has been run many times in the past. The alarm "Cutter Compensation Interference" comes up as an internal slot is being machined. What is the most likely cause of this alarm?

A. The machinist did not compensate for the edge finder during setup.

B. A four-flute endmill was installed instead of a two-flute endmill.

C. The tool length offset is incorrect.

D. The tool radius offset is incorrect.

22. After running a setup piece for the part in the following image, the 2.500″ dimension measures 2.496″ and the 1.250″ dimension measures 1.246″. What action is needed to correct this before running the next part?

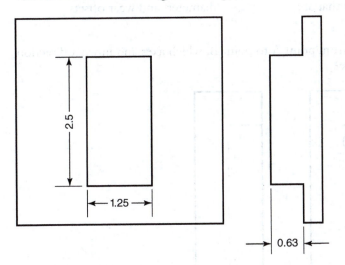

A. Adjust the radius/diameter wear offset by + 0.002″.

B. Adjust the tool radius offset of 0.500″ to 0.502″.

C. Replace the endmill.

D. Adjust the radius/diameter wear offset by 0.004″.

E. Either A or B is correct.

23. The threaded stud that is attached to a machining center's tool holder(s) and is gripped by a mechanism that secures the tapered tool holder in the spindle defines:

A. retention knob.

B. flange stud.

C. toggle clamp.

D. quick connect.

24. Incremental programming or specifying a distance from the current position instead of from the origin defines which code?

A. G98

B. G90

C. G91

D. G20

25. Which of the methods listed here is recommended when adding fluid to a coolant tank?

A. Add water until the desired level is reached.

B. Mix a 50:50 soluble oil to water mixture and add it to the tank.

C. Mix a 20:80 soluble oil to water mixture and add it to the tank.

D. Check the coolant ratio with a refractometer. Add soluble oil and/or water as necessary to obtain the desired concentration.

26. Select the answer that **best** describes the edit mode function.

    A. Allows programs to be modified

    B. Allows manual control of the axes movement

    C. Provides the screen to input and execute single commands

    D. The mode function that is selected when a program is running

27. Select the answer that **best** describes the offset page.

    A. Screen displaying settings and parameters that are turned on and off frequently

    B. List of machine parameters that are seldom changed

    C. Screen displaying work and machine coordinates as well as distance to go

    D. Screen displaying tool length, diameter, and wear offsets

28. To program the spindle to cut from point A to point B, which axes and in what directions would the spindle need to move?

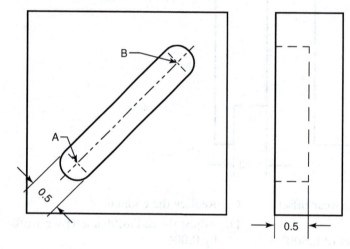

    A. X+ Y+

    B. X+ Y−

    C. X− Y−

    D. X− Y+

29. Referring to Geometric Dimensioning and Tolerancing, which of the following **best** describes a basic dimension?

    A. A numerical value used to describe a theoretical dimension has a box around the dimension

    B. A plane from which dimensions are referenced by a bold capital letter with a square box around it

    C. A rectangular box with a series of compartments for symbols and dimensions that describe a part feature

    D. None of the above

30. After milling the circular pockets, both diameters measure 0.007″ undersize. What offset adjustment will correct this when the tool is run again?

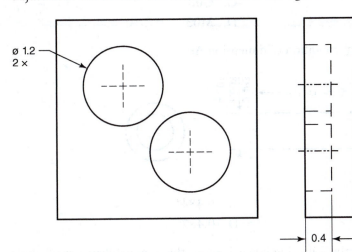

ø 1.2
2 ×

0.4

   A. +0.007″   C. −0.0035″

   B. −0.007″   D. −0.002″

31. The Geometric Dimensioning and Tolerancing symbol ⊥ describes:

   A. perpendicularity.   C. straightness.

   B. flatness.   D. tangent.

32. Which code corresponds with rapid positioning?

   A. G00   C. G03

   B. M00   D. M03

33. Select the mode switch position used to return a machine's axes to their home position.

   A. Auto   C. Edit

   B. MDI   D. Zero return

34. The allowable deviation from the nominal dimension of a part feature (±0.010″, ±1/16″) describes a:

   A. feature.   C. position.

   B. datum.   D. tolerance.

35. The formula RPM = 3.83 × SFPM/Ø is used to determine:

   A. feed rate.   C. pitch.

   B. torque.   D. spindle speed.

36. Which symbol below describes flatness?

   A. ÷   C. —

   B. //   D. ▱

37. To start the spindle on a CNC machine in the clockwise direction is initiated by what code?

    A. M04

    B. G04

    C. G03

    D. M03

38. What is the maximum material condition of dimension A?

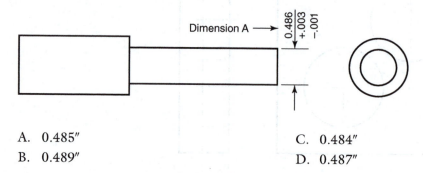

    A. 0.485″

    B. 0.489″

    C. 0.484″

    D. 0.487″

39. The zero on the bore gage in the following image is set to the nominal diameter of a hole. Based on the measurement in the following image, what adjustment to the boring bar is necessary to machine the bore to the nominal size on the next piece?

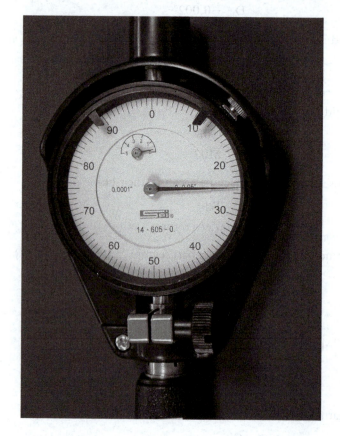

    A. −0.0002″

    B. +0.0025″

    C. −0.0025″

    D. −0.025″

40. What is the measurement of the 1–2″-depth micrometer in the following image?

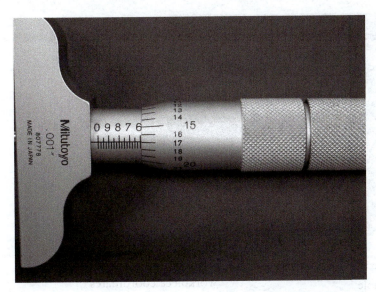

A. 1.592″

B. 1.567″

C. 2.617″

D. None of these

Continue with questions 41–70 if you are preparing for the CNC milling programming, setup, and operations exam.

41. The part print calls for a 0.500″ × 0.750″–deep hole to be drilled using a standard 118° twist drill. Approximately, how much past the 0.750″ depth must the hole be drilled to compensate for the tip length of the drill?

A. 0.250″

B. 0.150″

C. 0.075″

D. 0.050″

42. The material requirements for a workpiece specify 310 stainless steel with a Brinell hardness of 155. What RPM should the machine be programmed to for drilling through this part with a letter J drill? (Use 3.82 as the constant.)

A. 53

B. 965

C. 690

D. 722

43. An estimator is calculating the cycle time of a production part. One of the holes on this part is a 1/2″ Ø hole through 4.5″-thick cold rolled steel. The RPMs are set at 520 and the feed rate is 0.004 IPR. How many minutes will it take to drill through this part?

A. 2.16 minutes

B. 4.0 minutes

C. 5.20 minutes

D. 8.0 minutes

44. The tool in the following image is a(n) _____ and is used to _____.

    A. counterbore; machine a circular recess

    B. pin gage; measure a hole

    C. edge finder; establish x and y workpiece coordinates

    D. centering gage; establish tool offsets

45. The origins for both workpieces in the following image are in the upper-left corner of each workpiece. What coordinates should be entered in the offset page if the workpiece in vise A is to be machined?

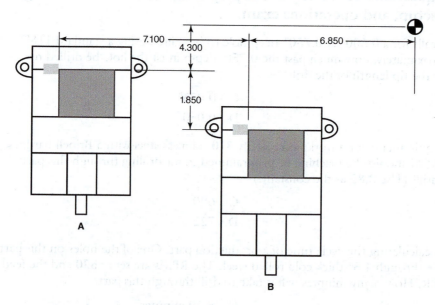

    A. X−6.85, Y−6.15

    B. X13.95, Y4.30

    C. X−13.95, Y−4.30

    D. X7.10, Y4.30

46. When setting up a vertical machining center, which technique(s) listed here is/are safe and acceptable?

    A. Use only two hold-down clamps.

    B. When the workpiece is large, position it on parallels and secure it with hold-down clamps, but do not position the clamps near the parallels.

    C. Position two or more clamps at opposite locations so the clamping force is evenly distributed.

    D. One clamp will work if the piece is thicker than 2.0″.

47. When using tool length compensation, the _____ word activates the corresponding tool length offset from the offset registry.

    A. H

    B. D

    C. Z

    D. P

48. A machinist is indicating the vise parallel to the *x*-axis of the machine. Which part of the vise should be used as the reference surface?

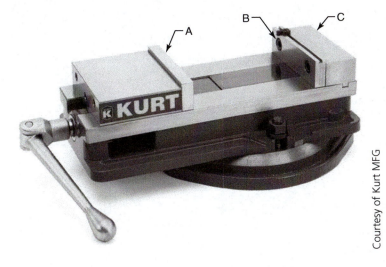

Courtesy of Kurt MFG

    A. A

    B. B

    C. C

    D. Either A or B

Refer to the program and the part print in the following figures to answer questions 49–60.

```
%
O8899
N5 G20 G90 G40 G49 G80 G94;
N10 M06 T1 (#3 CENTERDRILL);
N15 G0 G43 H1 Z8.0;
N20 G0 X0.0 Y0.0 G54;
N25 X.88 Y.75;
N30 M03 S3000;
N35 G0 Z3.0;
N40 Z.200 M08;
N45 G__ Z-.180 R.200 F12.0;
N50 X1.75 Y2.63;
N55 X2.63 Y2.25;
N60 G__ M09;
N65 G0 Z8.0;
N70 M06 T2 (5/16" DRILL);
N75 G0 G43 H2 Z8.0;
N80 M03 S3056;
N85 G0 X___ Y____;
N90 Z3.0;
N95 Z .200;
N100 G__ Z-____ R.200 Q.090 F15.00;
N105 G80 M9;
N110 G0 Z8.0;
N115 M06 T3 (LETTER X DRILL);
N120 G0 G43 H3 Z8.0;
N125 M03 S_____;
N130 G0 X1.75 Y1.5;
N135 Z3.0;
N140 Z.200 M08;
N145 G83 Z-.870 R.200 Q.100 F15.0;
N150 G80 M09;
N155 G0 Z 8.0;
N160 M6 T4 (1/2" COUNTERSINK);
N165 G0 G43 H4 Z8.0;
N170 M03 S1140;
N175 G0 X1.75 Y1.5;
N180 Z3.0;
N185 Z.200 M08;
N190 G82 Z-.215 R.200 P3000 F10.0;
N195 G80;
N200 G0 Z8.0 M09;
N205 M6 T5 (3/8-16 TAP);
N210 G0 G43 H5 Z8.0;
N215 M03 S470;
N220 G0 X.88 Y.75;
N225 Z3.0;
N230 Z.200 M08;
N235 G43 Z.050 H5;
N240 G__ X.88 Y.750 R.050 Z-.8 F_____;
N245 G80;
N250 M09;
N255 G0 Z8.0;
```

Letter X
Drill Thru
0.437 ø Chamfer

0.6250 Ream Thru

3/8–16 UNC 2B
Thru

2.25

1.5

0.75

0.88

1.75

2.63

3.5

49. The M03 code on line N30 commands the program to:

     A. program end.   C. spindle on CW.
     B. spindle on CCW.   D. spindle stop.

50. The M08 code in line N40 commands the machine to:

     A. coolant on.   C. coolant off.
     B. spindle on CW.   D. spindle stop.

51. To spot drill the three holes on line N45, what G code should be used?

     A. G01   C. G81
     B. G80   D. G83

52. To cancel a canned cycle on line N60, what code should be used?

     A. M0   C. G90
     B. M01   D. G80

53. On line N85, the x and y coordinates for the threaded hole should be:

     A. X.75 Y1.75.   C. X.88 Y.75.
     B. X1.75 Y1.5.   D. X2.25 Y2.63.

54. The code that will command the machine for a deep hole peck drilling cycle on line N100 is:

     A. G81.   C. G83.
     B. G76.   D. G41.

55. The finished depth of the threaded hole that must go completely through the part is indicated on line N100 by (the part is 0.750″ thick):

     A. Z−.750.   C. Z−.775.
     B. Z−.844.   D. None of the above

56. On line N125, the calculated RPM value should be (use 3.82 for the constant and 350 SFPM):

     A. 3368.   C. 3300.
     B. 2500.   D. 3056.

57. The depth of the countersink tool on line N190 is controlled by:

     A. Z−.215.   C. Z−.250.
     B. Z−.397.   D. Z−.350.

58. The P3000 code on line N195 indicates:

     A. dwell for 3 seconds.   C. speed of 3000 RPM.
     B. peck 30 times.   D. None of the above.

59. To command a canned tapping cycle on line N240, which code should be used?

     A. G83   C. G85
     B. G84   D. G86

60. The feed rate (IPM) that will need to be applied on line N240 to ensure that the 3/8-16 tap will cut at the correct rate is:

A. F15.620.

B. F20.

C. F.006.

D. F29.375.

Refer to the program and the part print in the following figures to answer questions 61–70.

```
%
O8900
N5 G20 G90 G40 G49 G80;
N10 M06 T1 (.500 ENDMILL);
N15 G0 Z8.0 G43 H1;
N20 G54 G0 X0.0 Y0.0;
N25 M03 S2400;
N30 G0 X0.0 Y-1.0;
N35 Z1.0;
N40 Z.100;
N45 G__ Z-.300 F15.0;
N50 X0.0 Y-.5 G41 D1;
N55 X0.0 Y.6;
N60 X___ Y____(Point 5);
N65 G__ X_____ Y____ I____ J_____ F15.0;
N70 G01 X0.0 Y1.4;
N75 X0.0 Y1.75;
N80 X_____ Y____(Point 9);
N85 X1.872 Y2.0(Point 10);
N90 G__ X_____ Y____ I____ J____ F15.0;
N95 G01 X_____ Y____ F15.0(Point 12);
N100 X1.34 Y0.0;
N105 X-.5 Y0.0;
N110 G40 G0 Y-1.0;
N115 Z8.0 M9;
N120 M5;
N125 M__;
```

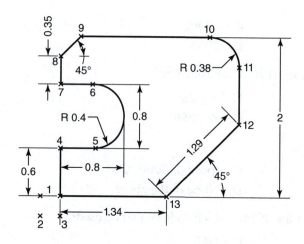

61. On line N45, select the G code for linear interpolation.

    A. M00   C. G01

    B. G02   D. G00

62. On line N50, what will the G41 command the machine to do?

    A. Dwell   C. Tool length cancel

    B. Workpiece coordinate select   D. Cutter compensation left

63. On line N60, what are the *x* and *y* coordinates for position 5?

    A. X.400, Y.600   C. X.600, Y.800

    B. X.800, Y.600   D. X.500, Y.500

64. Which line of code will complete the arc on line 65?

    A. G02 X.4 Y1.4 I.4 J.4 F15.0   C. G02 X.6 Y.8 I0.0 J.4 F15.0

    B. G03 X.4 Y1.4 I0.0 J.4 F15.0   D. G03 X.4 Y1.4 I.4 J0.0 F15.0

65. What X and Y coordinates will correspond to point 9 on line N80?

    A. X0.0, Y1.75   C. X.25, Y2.0

    B. X.35, Y1.75   D. X.35, Y1.75

66. Which line of code will complete the arc on line N90?

    A. G02 X1.72 Y1.65 I0.0 J−.35 F15.0   C. G02 X2.252 Y1.62 I0.0 J−.38 F15.0

    B. G03 X2.0 Y1.62 I.38 J−.38 F15.0   D. G03 X2.250 Y1.62 I0.0 J−.38 F15.0

67. What are the X and Y coordinates of point 12 on line N95?

    A. X2.63, Y1.29   C. X1.87, Y1.0

    B. X2.252, Y.912   D. X2.0, Y.912

68. What will the G40 code on line N110 command the machine to do?

    A. Dwell   C. Coolant off

    B. Cancel the tool length offset   D. Cutter compensation cancel

69. On line N125, what code will end the program and return to the start of the program?

    A. M9   C. M30

    B. M02   D. G03

70. On line N90, to machine the radius from point 10 to point 11 using an R value instead of *I* and *J*, what line of code should be used?

    A. G03 X2.250 Y1.62 R.38   C. G02 X2.252 Y1.62 R.38

    B. G03 X2.250 Y1.62 R.38   D. G02 X2.250 Y1.62 R-−.38

8.2 ANSWER KEY

1.	D	26.	A	51.	C
2.	C	27.	D	52.	D
3.	E	28.	A	53.	C
4.	C	29.	A	54.	C
5.	C	30.	C	55.	B
6.	C	31.	A	56.	A
7.	D	32.	A	57.	A
8.	B	33.	D	58.	A
9.	C	34.	D	59.	B
10.	B	35.	D	60.	D
11.	D	36.	D	61.	C
12.	D	37.	D	62.	D
13.	C	38.	B	63.	A
14.	B	39.	C	64.	B
15.	D	40.	A	65.	C
16.	B	41.	B	66.	C
17.	B	42.	C	67.	B
18.	A	43.	A	68.	D
19.	C	44.	C	69.	C
20.	D	45.	C	70.	C
21.	D	46.	C		
22.	E	47.	A		
23.	A	48.	B		
24.	C	49.	C		
25.	D	50.	C		

8.3 PRACTICE THEORY EXAM EXPLANATIONS

1. Which code activates the workpiece coordinate system?

   A. G43
   B. G41
   C. G17
   D. G54

   Answer D is correct. G54 is the first offset number on a machining center's work offset page. Calling up this offset in a program will provide the MCU with distance from machine zero to the workpiece zero.

2. Which code directs the program to end and return to start?

   A. M02
   B. M99
   C. M30
   D. G80

   Answer C is correct. M30 is the M code that will end the program and reset the program back to the beginning.

3. Which of the following answers **best** identifies a countersink tool?

   A. A tool used to produce a flat bearing surface on a rough, uneven surface
   B. A tool used to create an angled bevel to remove burrs, aid the start of taps, and the entry of dowel pins
   C. A tool used to cut a larger-diameter hole to a specific depth to allow a bolt head or nut to sit flush or below the part's surface
   D. A tool used to cut a tapered opening at the beginning of an existing hole to a specific depth to allow a flathead screw to sit flush or below the part's surface
   E. B and D are correct

   Answer E is correct. A countersink is a cutting tool used to machine chamfers on the opening of a hole and to machine a conical opening at the beginning of a hole to receive the head of a flathead screw.

4. Typically done as part of the start-up procedure, which is the process of positioning the CNC machining center's axes at a fixed reference position to allow the MCU to accurately establish/track the table's position?

   A. Jogging
   B. Calibrating
   C. Homing
   D. Positioning

   Answer C is correct. Homing is the act of positioning the machine's axes at their reference position, also called machine zero position. Once at this position, information is relayed back to the MCU and the coordinates are verified. This is typically accomplished automatically by pressing a button(s) or key(s) on the machine's control panel.

5. To utilize an M01 command, which mode function would have to be selected?

   A. Block delete
   B. Select program
   C. Optional stop
   D. Single block

   Answer C is correct. The M01 code will stop the program when the optional stop button on the control is enabled.

6. After roughing an internal pocket, a program stop is utilized in the program so the metal chips can be cleared. What will restart the machine from this point in the program?

    A. Power up/restart   C. Cycle start

    B. Power on   D. Reset

    Answer C is correct. The cycle start button will restart the program at the line after the M00 code.

7. Which of the following describes a material safety data sheet (MSDS)?

    A. MSDS is used to comply with OSHA's hazard communication standard.

    B. It identifies the chemicals in a workplace and their recommended uses.

    C. It provides information about safe handling of chemicals and materials.

    D. All of the above describe an MSDS.

    Answer D is correct. Material safety data sheets are utilized by manufacturers of chemicals and materials to identify chemicals and their uses, communicate hazards, and provide recommendations for safe storage and handling.

8. Select the code to fill in the blank. To change tools on a vertical machining center from the tool currently in the spindle to tool number 8 using MDI, the machinist would input _____ T8.

    A. G96   C. M08

    B. M06   D. M02

    Answer B is correct. Calling the M06 code with a tool number will command the machine to change the tool in the spindle to the tool number specified.

9. A part print calls for a workpiece to have a finished length of 2.025″. Leaving 0.100″ on the length for machining operations, what is the length of stock that must be cut?

    A. 2 1/16″   C. 2 1/8″

    B. 2 3/32″   D. 2 1/4″

    Answer C is correct. 2.025″ + 0.100″ = 2.125″, which is the decimal equivalent of 2 1/8″.

10. A machinist has just set up a VMC and ran a setup piece for a program that has been proven and run in the past. The print in the following image shows the desired hole pattern and the actual hole pattern, which was machined on the setup piece. What is the likely cause of this situation?

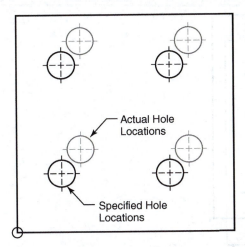

A. The program coordinates are incorrect.

B. The work coordinates or fixture offset is incorrect.

C. The tool offset is incorrect.

D. There is an M30 in the program.

Answer B is correct. The program was run in the past, which indicates that the geometry and locations are correct in the program. When all of the part's features are incorrectly located an equal distance from their individual specified locations, the cause will typically be an error in the workpiece coordinates.

11. After milling the pocket in the following image, the depth measurement is 0.747″. Which of the following answers will **correctly** adjust this depth when this tool in the program is run again?

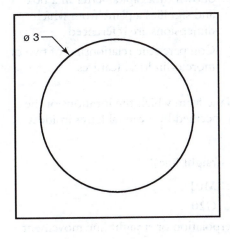 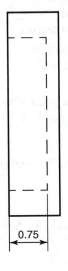

A. Change the Z value for this tool in the program.

B. Adjust the Z wear offset +.003″.

C. Place shims under the workpiece.

D. Adjust the Z wear offset −.003″.

E. All of the above

Answer D is correct. Adjusting the Z wear offset by a value of −.003″ will command the MCU to move this tool past its programmed depth by 0.003″.

12. You are running a setup piece to produce the part in the following image. The diameter of the pocket finishes at 2.992″. What change to the wear offset will be necessary to machine this pocket to the nominal dimension?

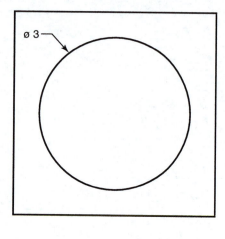
ø 3

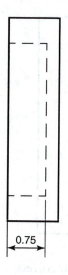
0.75

A. +.005″

B. +.004″

C. −.006″

D. −.004″

Answer D is correct. The measured dimension of 2.992″ is 0.008″ under the nominal dimension. Adjusting the radius offset for this tool in the wear offset page will command this tool to cut closer to the nominal size of the circle by 0.004″ per side, or 0.008″ on the diameter.

13. Which statement **best** describes a datum?

A. A rectangular box with a series of compartments for symbols and dimensions that describe a part feature

B. A numerical value positioned within a box used to describe a theoretical dimension

C. Shown by a capital letter in a box and signifies a plane from which dimensions are referenced

D. Compares the relationship of two or more cylindrical features

Answer C is correct. A datum is a point, axis, or plane from which the locations of the part's geometric features are referenced. A datum is specified by a capital letter inside a square box.

14. Which code identifies linear interpolation (feed in a straight line)?

A. G00

B. G01

C. M01

D. G20

Answer B is correct. G01 is the code for linear interpolation or straight line movement controlled by a feed rate.

15. In which of the following scenarios is the individual following recommended safety guidelines?

    A. A machine tool service technician is repairing the machine you operate and is only wearing prescription glasses with no side shields

    B. The plant manager is watching from the aisle of the shop and is not wearing eye protection

    C. A coworker puts approved safety glasses on while the tool is cutting and then removes them to measure the workpiece

    D. You wear safety glasses every time you enter the shop floor and remove them only when you have exited the manufacturing environment

Answer D is correct. Approved eye protection should be worn by everyone at any time he or she is on the manufacturing floor.

16. To maintain proper coolant levels and concentrations and to ensure a machine tool has sufficient lubrication, a machinist should check the machine's fluids:

    A. at the end of each week.

    B. every shift.

    C. every other week.

    D. at the beginning of each week.

Answer B is correct. Every machine tool manufacturer has its own recommendations for maintenance intervals and procedures. To avoid unnecessary wear to the machining center and to maintain optimal cutting conditions, the coolant and lubrication system should be checked at the start of every shift.

17. Select the answer that **best** describes the Jog mode function.

    A. Allows programs to be modified

    B. Allows manual control of axes movement

    C. Sends the machine to its reference position

    D. Must be selected to run a program

Answer B is correct. Manual control of the machine's axes through the use of buttons or a rotary handwheel is accomplished in the jog mode.

18. Select the answer that **best** describes the position page on a CNC control.

    A. A digital reference screen that displays work and machine coordinates as well as the distance to go for a positioning command

    B. A screen displaying tool geometry, wear, and work offsets

    C. Lists machine parameters that are seldom changed

    D. Displays settings and parameters that are turned on and off frequently

Answer A is correct. The position page provides the operator with position data for each axis of the machine. The position page on a typical CNC control will display coordinates in the following categories: position-operator, position-work, position-machine, and distance-to-go position.

19. If you are standing in front of a horizontal CNC milling machine, which axis moves in and out (front to back)?

    A. x

    B. y

    C. z

    D. C

Answer C is correct. The horizontal machining center gets its name from the horizontal orientation of the spindle (z-axis). When facing the front of the machine, this axis moves in and out, toward and away from the operator.

20. G codes are referred to as:

    A. miscellaneous commands.   C. position codes.

    B. canned cycles.   D. preparatory commands.

    Answer D is correct. Codes that instruct the machine to perform machining operations and/or input machine settings are called preparatory commands.

21. A machinist is running a setup piece for a part program that has been run many times in the past. The alarm "Cutter Compensation Interference" comes up as an internal slot is being machined. What is the most likely cause of this alarm?

    A. The machinist did not compensate for the edge finder during setup.   C. The tool length offset is incorrect.

    D. The tool radius offset is incorrect.

    B. A four-flute endmill was installed instead of a two-flute endmill.

    Answer D is correct. For the tool to move from a center coordinate tool path to a tangent coordinate tool path, a lead-in move must be executed. If a distance of at least the radius of the tool is not available for this move, an alarm will most likely occur. In this example, the actual diameter/radius of the tool may be slightly larger than the lead-in distance.

22. After running a setup piece for the part in the following image, the 2.500″ dimension measures 2.496″ and the 1.250″ dimension measures 1.246″. What action is needed to correct this before running the next part?

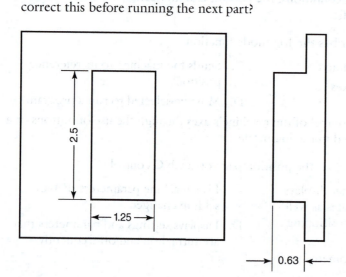

    A. Adjust the radius/diameter wear offset by +.002″.   C. Replace the endmill.

    D. Adjust the radius/diameter wear offset by 0.004″.

    B. Adjust the tool radius offset of 0.500″ to 0.502″.

    E. Either A or B is correct.

    Answer E is correct. The tool used to cut this profile must be adjusted to leave an additional 0.002″ per surface. The preferred method is to add 0.002 to the wear offset for this tool. If the control is older and does not have a wear offset page, the tool radius value can be increased by 0.002.

23. The threaded stud that is attached to a machining center's tool holder(s) and is gripped by a mechanism that secures the tapered tool holder in the spindle defines:

    A. retention knob.

    B. flange stud.

    C. toggle clamp.

    D. quick connect.

    Answer A is correct. A retention knob or pull stud threads into the tapered end of a toolholder. A mechanical mechanism in the machine's spindle slides under the knob portion and draws the toolholder in the spindle's taper.

24. Incremental programming or specifying a distance from the current position instead of from the origin defines which code?

    A. G98

    B. G90

    C. G91

    D. G20

    Answer C is correct. G91 is the preparatory command for activating incremental positioning.

25. Which of the methods listed here is recommended when adding fluid to a coolant tank?

    A. Add water until the desired level is reached.

    B. Mix a 50:50 soluble oil to water mixture and add it to the tank.

    C. Mix a 20:80 soluble oil to water mixture and add it to the tank.

    D. Check the coolant ratio with a refractometer. Add soluble oil and/or water as necessary to obtain the desired concentration.

    Answer D is correct. To provide the maximum cooling and lubricity, water-soluble cutting fluids must be properly maintained. The concentration level should be checked with a refractometer and adjusted according to the manufacturer's recommendations.

26. Select the answer that **best** describes the edit mode function.

    A. Allows programs to be modified

    B. Allows manual control of the axes movement

    C. Provides the screen to input and execute single commands

    D. The mode function that is selected when a program is running

    Answer A is correct. The edit mode on a CNC control is the mode where new programs can be typed into the control, existing programs can be modified, and programs can be installed prior to use.

27. Select the answer that **best** describes the offset page.

    A. Screen displaying settings and parameters that are turned on and off frequently

    B. List of machine parameters that are seldom changed

    C. Screen displaying work and machine coordinates as well as distance to go

    D. Screen displaying tool length, diameter, and wear offsets

    Answer D is correct. The offset page on a typical CNC control consists of three screens to store offset-related data. The three pages are the geometry offsets, wear offsets, and workpiece offsets.

28. To program the spindle to cut from point A to point B, which axes and in what directions would the spindle need to move?

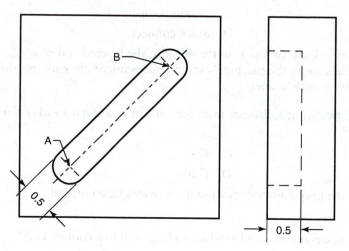

A. X+ Y+

B. X+ Y–

C. X– Y–

D. X– Y+

Answer A is correct. To machine this slot moving from point A to point B requires simultaneous movement of the *x*- and *y*-axes. The tool would cut along the *x*-axis from left to right, which is movement in the positive direction. The tool would also cut along the *y*-axis moving from bottom to top, which is movement in the positive direction for this axis.

29. Referring to Geometric Dimensioning and Tolerancing, which of the following **best** describes a basic dimension?

A. A numerical value used to describe a theoretical dimension has a box around the dimension

B. A plane from which dimensions are referenced by a bold capital letter with a square box around it

C. A rectangular box with a series of compartments for symbols and dimensions that describe a part feature

D. None of the above

Answer A is correct. A basic dimension is represented by a boxed symbol and is a theoretical precise value used to designate the exact size or location of a geometric feature.

30. After milling the circular pockets, both diameters measure 0.007″ undersize. What offset adjustment will correct this when the tool is run again?

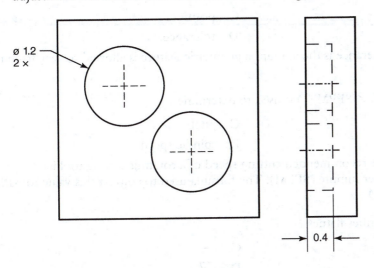

ø 1.2
2 ×

0.4

A. +.007″   C. −.0035″
B. −.007″   D. −.002″

Answer C is correct. To adjust the tool path so the pocket is cut to the nominal diameter, the wear offset for this tool should be adjusted. Adjusting the radius offset for this tool in the wear offset page by −0.0035 will command this tool to cut closer to the nominal size of the circles by 0.0035″ per side, or 0.007″ on the diameter.

31. The Geometric Dimensioning and Tolerancing symbol ⊥ describes:

A. perpendicularity.   C. straightness.
B. flatness.   D. tangent.

Answer A is correct. The Geometric Dimensioning and Tolerancing symbol ⊥ specifies that a geometric feature must be 90° to a specified datum.

32. Which code corresponds with rapid positioning?

A. G00   C. G03
B. M00   D. M03

Answer A is correct. Rapid traverse or rapid positioning occurs when a machining center's axes are moved quickly to a specified position. This is accomplished in a program with a G00 code.

33. Select the mode switch position used to return a machine's axes to their home position.

A. Auto   C. Edit
B. MDI   D. Zero return

Answer D is correct. Zero return will position the machine's axes at their reference position or machine zero position.

34. The allowable deviation from the nominal dimension of a part feature ($\pm.010''$, $\pm 1/16''$) describes a:

   A. feature.
   B. datum.
   C. position.
   D. tolerance.

   Answer D is correct. Tolerance is the amount a geometric feature is allowed to vary from its specified nominal size.

35. The formula RPM = $3.83 \times$ SFPM/Ø is used to determine:

   A. feed rate.
   B. torque.
   C. pitch.
   D. spindle speed.

   Answer D is correct. The recommended cutting speed of a rotating cutting tool is expressed in surface feet per minute (SFPM). The formula used to convert this value to RPM is RPM=$3.83 \times$ SFPM/Ø.

36. Which symbol below describes flatness?

   A. $\div$
   B. //
   C. —
   D. $\Box$

   Answer D is correct. The Geometric Dimensioning and Tolerancing symbol $\Box$ indicates that a specified surface must be flat within a specified tolerance.

37. To start the spindle on a CNC machine in the clockwise direction is initiated by what code?

   A. M04
   B. G04
   C. G03
   D. M03

   Answer D is correct. M03 is the code to start the spindle in the clockwise direction.

38. What is the maximum material condition of dimension A?

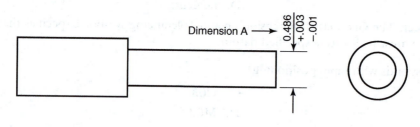

   A. 0.485''
   B. 0.489''
   C. 0.484''
   D. 0.487''

   Answer B is correct. The maximum material condition or MMC would occur when the part is within its specified tolerance and contains the most material mass. This would occur when the shaft is at the upper limit of the tolerance.

39. The zero on the bore gage in the following image is set to the nominal diameter of a hole. Based on the measurement in the following image, what adjustment to the boring bar is necessary to machine the bore to the nominal size on the next piece?

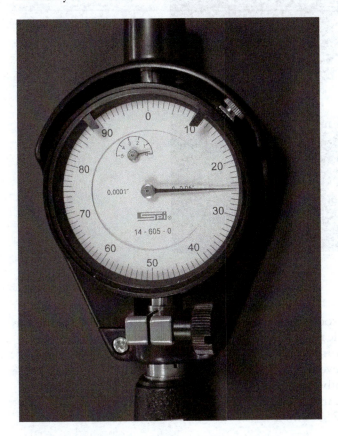

A. −0.0002″

B. +0.0025″

C. −0.0025″

D. −0.025″

Answer C is correct. Each graduation on the dial bore gage is 0.0001″. The needle is 25 graduations or 0.0025″ past the zero. The boring bar would have to be adjusted −0.0025″ to cut to the nominal size.

40. What is the measurement of the 1–2″-depth micrometer in the following image?

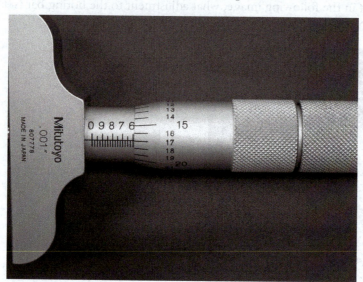

A. 1.592″

B. 1.567″

C. 2.617″

D. None of these

Answer A is correct. The 1–2″ rod is installed in the depth micrometer, which indicates that the measurement will be at least 1.0″ deep. The 5 on the sleeve of the depth micrometer is not visible, indicating 0.500″ is added to the reading. The three vertical lines (after the 5 line) on the sleeve are also not visible, indicating 0.075″ is added to the reading. The horizontal line numbered 17 on the thimble aligns to the horizontal line running the length of the sleeve, indicating 0.017″ is added to the reading. 1.0″ + 0.500″ + 0.075″ + 0.017″ = 1.592″.

Continue with questions 41–70 if you are preparing for the CNC milling programming, setup, and operations exam.

41. The part print calls for a 0.500″ × 0.750″–deep hole to be drilled using a standard 118° twist drill. Approximately, how much past the 0.750″ depth must the hole be drilled to compensate for the tip length of the drill?

A. 0.250″

B. 0.150″

C. 0.075″

D. 0.050″

Answer B is correct. As a general rule of thumb, to calculate the length of a drill point on a standard 118° drill, multiply the diameter of the drill by 0.300.   0.300 × 0.500″ = 0.150″.

42. The material requirements for a workpiece specify 310 stainless steel with a Brinell hardness of 155. What RPM should the machine be programmed to for drilling through this part with a letter J drill? (Use 3.82 as the constant.)

A. 53

B. 965

C. 690

D. 722

Answer C is correct. Inserting the provided variables into the RPM formula will result in

$$\text{RPM} = \frac{3.82 \times 50}{0.277} = 689.53.$$

43. An estimator is calculating the cycle time of a production part. One of the holes on this part is a 1/2″ Ø hole through 4.5″-thick cold rolled steel. The RPMs are set at 520 and the feed rate is 0.004 IPR. How many minutes will it take to drill through this part?

A. 2.16 minutes

B. 4.0 minutes

C. 5.20 minutes

D. 8.0 minutes

Answer A is correct. The time is calculated by $\dfrac{4.5}{520 \times 0.004} = 2.16.$

44. The tool in the following image is a(n) _____ and is used to _____.

A. counterbore; machine a circular recess

B. pin gage; measure a hole

C. edge finder; establish *x* and *y* workpiece coordinates

D. centering gage; establish tool offsets

Answer C is correct. When locating the *x* and *y* origins on square parts with edges parallel to the *x*- and *y*-axes of the machine, an edge finder may be used.

45. The origins for both workpieces in the following image are in the upper-left corner of each workpiece. What coordinates should be entered in the offset page if the workpiece in vise A is to be machined?

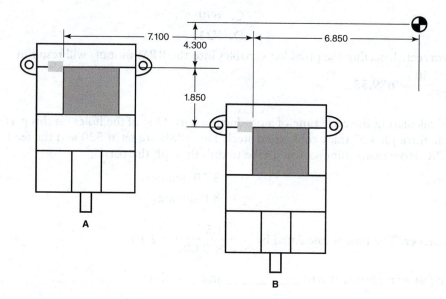

A. X−6.85, Y−6.15

B. X13.95, Y4.30

C. X−13.95, Y−4.30

D. X7.10, Y4.30

Answer C is correct. The distance from the machine's home position on the *x*-axis is −13.95″ and on the *y*-axis is −4.30″.

46. When setting up a vertical machining center, which technique(s) listed here is/are safe and acceptable?

A. Use only two hold-down clamps.

B. When the workpiece is large, position it on parallels and secure it with hold-down clamps, but do not position the clamps near the parallels.

C. Position two or more clamps at opposite locations so the clamping force is evenly distributed.

D. One clamp will work if the piece is thicker than 2.0″.

Answer C is correct. Regardless of the type of clamp, a minimum of two clamping points (more if feasible) with the force applied directly down on the workpiece should be utilized.

47. When using tool length compensation, the _____ word activates the corresponding tool length offset from the offset registry.

A. H

B. D

C. Z

D. P

Answer A is correct. The "H" word followed by a number that corresponds to the tool being used will indicate to the control what tool length offset to apply.

48. A machinist is indicating the vise parallel to the *x*-axis of the machine. Which part of the vise should be used as the reference surface?

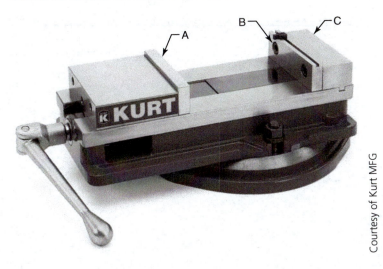

Courtesy of Kurt MFG

A. A

B. B

C. C

D. Either A or B

Answer B is correct. The solid, stationary jaw should be used when indicating a vise. Because of its fixed position, it will not produce the variation the movable jaw is likely to produce.

Refer to the program and the part print in the following figures to answer questions 49–60.

```
%
O8899
N5 G20 G90 G40 G49 G80 G94;
N10 M06 T1 (#3 CENTERDRILL);
N15 G0 G43 H1 Z8.0;
N20 G0 X0.0 Y0.0 G54;
N25 X.88 Y.75;
N30 M03 S3000;
N35 G0 Z3.0;
N40 Z.200 M08;
N45 G__ Z-.180 R.200 F12.0;
N50 X1.75 Y2.63;
N55 X2.63 Y2.25;
N60 G__ M09;
N65 G0 Z8.0;
N70 M06 T2 (5/16" DRILL);
N75 G0 G43 H2 Z8.0;
N80 M03 S3056;
N85 G0 X___ Y____;
N90 Z3.0;
N95 Z .200;
N100 G__ Z-____ R.200 Q.090 F15.00;
N105 G80 M9;
N110 G0 Z8.0;
N115 M06 T3 (LETTER X DRILL);
N120 G0 G43 H3 Z8.0;
N125 M03 S_____;
N130 G0 X1.75 Y1.5;
N135 Z3.0;
N140 Z.200 M08;
N145 G83 Z-.870 R.200 Q.100 F15.0;
N150 G80 M09;
N155 G0 Z 8.0;
N160 M6 T4 (1/2" COUNTERSINK);
N165 G0 G43 H4 Z8.0;
N170 M03 S 1140;
N175 G0 X1.75 Y1.5;
N180 Z3.0;
N185 Z.200 M08;
N190 G82 Z-.215 R.200 P3000 F10.0;
N195 G80;
N200 G0 Z8.0 M09;
N205 M6 T5 (3/8-16 TAP);
N210 G0 G43 H5 Z8.0;
N215 M03 S470;
N220 G0 X.88 Y.75;
N225 Z3.0;
N230 Z.200 M08;
N235 G43 Z.050 H5;
N240 G__ X.88 Y.750 R.050 Z-.8 F_____;
N245 G80;
N250 M09;
N255 G0 Z8.0;
```

Letter X
Drill Thru
0.437 ø Chamfer

0.6250 Ream Thru

3/8–16 UNC 2B
Thru

2.25

1.5

0.75

0.88

1.75

2.63

3.5

49. The M03 code on line N30 commands the program to:

    A. program end.   C. spindle on CW.

    B. spindle on CCW.   D. spindle stop.

    Answer C is correct. M03 is the code to start the spindle in the clockwise direction.

50. The M08 code in line N40 commands the machine to:

    A. coolant on.   C. coolant off.

    B. spindle on CW.   D. spindle stop.

    Answer C is correct. M08 is the code for the coolant on command.

51. To spot drill the three holes on line N45, what G code should be used?

    A. G01   C. G81

    B. G80   D. G83

    Answer C is correct. The G81 canned drilling cycle is a standard drilling cycle to a specified depth.

52. To cancel a canned cycle on line N60, what code should be used?

    A. M0   C. G90

    B. M01   D. G80

    Answer D is correct. Because canned cycles are modal commands, a G80 code is used to cancel the canned cycle prior to proceeding to the next machining operation.

53. On line N85, the *X* and *Y* coordinates for the threaded hole should be:

    A. X.75 Y1.75.   C. X.88 Y.75.

    B. X1.75 Y1.5.   D. X2.25 Y2.63.

    Answer C is correct. The coordinates of the 3/8-16 UNC threaded hole are $X\,0.88$, $Y\,0.75$.

54. The code that will command the machine for a deep hole peck drilling cycle on line N100 is:

    A. G81.   C. G83.

    B. G76.   D. G41.

    Answer C is correct. The G83 canned drilling cycle is a standard peck drilling cycle to a specified depth.

55. The finished depth of the threaded hole that must go completely through the part is indicated on line N100 by (the part is 0.750″ thick):

    A. Z−.750.   C. Z−.775.

    B. Z−.844.   D. None of the above

    Answer B is correct. As a general rule of thumb, to calculate the length of a drill point on a standard 118° drill, multiply the diameter of the drill by 0.300.

    Example: $0.300 \times 0.3125 = 0.09375″$. Add 0.09375 to the 0.750″ part thickness and 0.844″ is the depth the drill should be programmed to.

56. On line N125, the calculated RPM value should be (use 3.82 for the constant and 350 SFPM):

A. 3368.

C. 3300.

B. 2500.

D. 3056.

Answer A is correct. To calculate the RPM value for this tool, use the formula:

$$RPM = \frac{3.82 \times 350}{0.397} = 3367.75$$

57. The depth of the countersink tool on line N190 is controlled by:

A. Z−.215.

C. Z−.250.

B. Z−.397.

D. Z−.350.

Answer A is correct. The G82 canned cycle is similar to a standard drilling cycle, but it has a built-in dwell. The Z value specifies the final depth of the drill.

58. The P3000 code on line N195 indicates:

A. dwell for 3 seconds.

C. speed of 3000 RPM.

B. peck 30 times.

D. None of the above

Answer A is correct. When using the canned drilling cycle with a dwell, the "P" word specifies how long the tool will pause or dwell. P3000 is interpreted as 3000 milliseconds or 3 seconds.

59. To command a canned tapping cycle on line N240, which code should be used?

A. G83

C. G85

B. G84

D. G86

Answer B is correct. The G84 canned tapping cycle is a standard tapping cycle to a specified depth.

60. The feed rate (IPM) that will need to be applied on line N240 to ensure that the 3/8-16 tap will cut at the correct rate is:

A. F15.620.

C. F.006.

B. F20.

D. F29.375.

Answer D is correct. Using the formula Feed = RPM/TPI, the tapping feed rate can be calculated. Feed = 470/16 = 29.375.

Refer to the program and the part print in the following figures to answer questions 61–70.

```
%
O8900
N5 G20 G90 G40 G49 G80;
N10 M06 T1 (.500 ENDMILL);
N15 G0 Z8.0 G43 H1;
N20 G54 G0 X0.0 Y0.0;
N25 M03 S2400;
N30 G0 X0.0 Y-1.0;
N35 Z1.0;
N40 Z.100;
N45 G__ Z-.300 F15.0;
N50 X0.0 Y-.5 G41 D1;
N55 X0.0 Y.6;
N60 X___ Y____(Point 5);
N65 G__ X_____ Y____ I____ J_____ F15.0;
N70 G01 X0.0 Y1.4;
N75 X0.0 Y1.75;
N80 X_____ Y_____(Point 9);
N85 X1.872 Y2.0(Point 10);
N90 G__ X_____ Y_____ I____ J_____ F15.0;
N95 G01 X_____ Y_____ F15.0(Point 12);
N100 X1.34 Y0.0;
N105 X-.5 Y0.0;
N110 G40 G0 Y-1.0;
N115 Z8.0 M9;
N120 M5;
N125 M__;
```

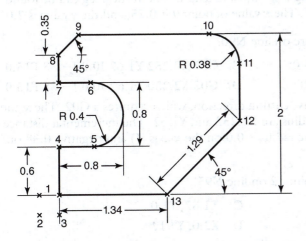

61. On line N45, select the G code for linear interpolation.

   A. M00

   B. G02

   C. G01

   D. G00

Answer C is correct. G01 is the code for linear interpolation or straight line movement controlled by a feed rate.

62. On line N50, what will the G41 command the machine to do?

    A. Dwell
    B. Workpiece coordinate select
    C. Tool length cancel
    D. Cutter compensation left

    Answer D is correct. G41 is the code for cutter compensation left. This code will command the tool to move to the left of the programmed tool path by an amount specified in the offset registry.

63. On line N60, what are the x and y coordinates for position 5?

    A. X.400, Y.600
    B. X.800, Y.600
    C. X.600, Y.800
    D. X.500, Y.500

    Answer A is correct. The coordinates of point 5 are x 0.400, y 0.600.

64. Which line of code will complete the arc on line 65?

    A. G02 X.4 Y1.4 I.4 J.4 F15.0
    B. G03 X.4 Y1.4 I0.0 J.4 F15.0
    C. G02 X.6 Y.8 I0.0 J.4 F15.0
    D. G03 X.4 Y1.4 I.4 J0.0 F15.0

    Answer B is correct. This is a counterclockwise cutting direction, which requires a G03. The x and y coordinates of the endpoint of the radius are X.4 and Y1.4. The incremental distance from the start point to the center of the radius is 0.0 on the x-axis (I) and positive 0.4 on the y-axis (J).

65. What x and y coordinates will correspond to point 9 on line N80?

    A. X0.0, Y1.75
    B. X.35, Y1.75
    C. X.25, Y2.0
    D. X.35, Y1.75

    Answer C is correct. Since the angle from point 8 to point 9 is 45°, the vertical and horizontal legs of the hypothetical triangle are equal in length. The vertical leg can be found by $2.0'' - 0.6'' - 0.8'' - 0.35'' = 0.25''$. The x value of point 9 is 0.25, and the y value is 2.0".

66. Which line of code will complete the arc on line N90?

    A. G02 X1.72 Y1.65 I0.0 J-.35 F15.0
    B. G03 X2.0 Y1.62 I.38 J-.38 F15.0
    C. G02 X2.252 Y1.62 I0.0 J-.38 F15.0
    D. G03 X2.250 Y1.62 I0.0 J-.38 F15.0

    Answer C is correct. This is a clockwise cutting direction, which requires a G02. The x and y coordinates of the endpoint of the radius are X2.252 and Y1.62. The incremental distance from the start point to the center of the radius is 0.0 on the x-axis (I) and negative 0.38 on the y-axis (J).

67. What are the x and y coordinates of point 12 on line N95?

    A. X2.63, Y1.29
    B. X2.252, Y.912
    C. X1.87, Y1.0
    D. X2.0, Y.912

    Answer B is correct. Since the angle created by the line extending from point 12 to point 13 is 45°, the horizontal and vertical legs of this hypothetical triangle are equal. The following equation can be used to calculate the length of these side of the triangle:

    $$\sin 45° = \frac{\text{Opposite}}{1.29} = 0.912.$$ To determine the x coordinate, add 0.912 to 1.34 = 2.252.

    To determine the r coordinate, add 0.912 to 0.0 = 0.912.

68. What will the G40 code on line N110 command the machine to do?

    A. Dwell
    B. Cancel the tool length offset
    C. Coolant off
    D. Cutter compensation cancel

    Answer D is correct. Cutter compensation is a modal command. A G40 code is used to cancel the cutter compensation prior to proceeding to the next machining operation.

69. On line N125, what code will end the program and return to the start of the program?

    A. M9
    B. M02
    C. M30
    D. G03

    Answer C is correct. M30 is the M code that will end the program and reset the program back to the beginning.

70. On line N90, to machine the radius from point 10 to point 11 using an *R* value instead of *I* and *J*, what line of code should be used?

    A. G03 X2.250 Y1.62 R.38
    B. G03 X2.250 Y1.62 R.38
    C. G02 X2.252 Y1.62 R.38
    D. G02 X2.250 Y1.62 R−.38

    Answer C is correct. This is a clockwise cutting direction, which requires a G02. The *x* and *y* coordinates of the endpoint of the radius are X2.252 and Y1.62. The *R* value is a positive 0.38 because the center of the radius lies outside the chord of the arc.

9 CNC Turning Certifications

9.1 PRACTICE THEORY EXAM CNC TURNING OPERATIONS/CNC TURNING PROGRAMMING, SETUP, AND OPERATIONS

Questions 1–40 are designed to prepare the student for the CNC operations exam and a portion of the CNC programming, setup, and operations exam. Questions 41–70 are designed to prepare the student for the additional concepts assessed by the CNC programming, setup, and operations exam.

1. Typically done as part of the start-up procedure, which of the following is the process of positioning the CNC turning center's axes at a fixed reference position to allow the MCU to accurately establish/track the machine's position?

    A. Jogging
    B. Homing
    C. Calibrating
    D. Positioning

2. To maintain proper coolant levels and concentrations, and to ensure a machine tool has sufficient lubrication, a machinist should check the machine's fluids:

    A. at the beginning of each week.
    B. at the end of each week.
    C. every shift.
    D. every other week.

3. Which code identifies linear interpolation (feed in a straight line)?

    A. G01
    B. G00
    C. M01
    D. G20

4. Your employer has appointed you to work with the safety committee to develop a department safety manual. Which safety rule listed here will help facilitate the safest work environment for the employees in this department?

    A. Store all tools in their appropriate location.
    B. Do not bypass door interlocks.
    C. Wear safety glasses at all times.
    D. Disconnect power sources when servicing a machine.
    E. Follow all safety rules at all times.

5. Referring to Geometric Dimensioning and Tolerancing, which of the following **best** describes a basic dimension?

 A. Title block tolerances are not applied to a basic dimension.

 B. A basic dimension is a numerical value used to describe a theoretical size, profile, orientation, or datum target.

 C. It has a box around the dimension.

 D. All of the above are correct.

 E. B and C only are correct.

6. Which statement **best** describes geometry offsets?

 A. Adjusted small amounts to compensate for tool wear and/or thermal variation

 B. Provides tool-specific data to the control, such as the length, nose radius, and tip position

 C. Sets the X and Z origin of the part

 D. Is accessed using G54

7. Which of the following tools is a right-hand turning tool and will work correctly with the M03 command when installed in a standard slant bed CNC lathe?

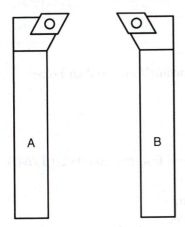

 A. Tool A fits this description.

 B. Tool B fits this description.

 C. Both tools fit this description.

 D. Neither tool fits this description.

8. Select the mode switch position used to return a machine's axes to their home position.

 A. Auto

 B. MDI

 C. Edit

 D. Zero return

9. Which of the methods listed here is recommended when adding fluid to a coolant tank?

 A. Add water until the desired level is reached.

 B. Mix a 50:50 soluble oil to water mixture and add it to the tank.

 C. Mix a 20:80 soluble oil to water mixture and add it to the tank.

 D. Check the coolant ratio with a refractometer. Add soluble oil and/or water as necessary to obtain the desired concentration.

10. A machinist needs to jog the tool from point A to point B in the following image. Which axis and in what direction should the tool be moved?

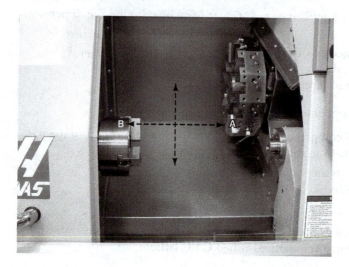

   A. *Z*; –

   B. *X*; –

   C. *Y*; –

   D. *X*; +

11. Which code directs a CNC Turning Center to move the machine's axes to their home positions?

   A. G28

   B. G20

   C. G90

   D. G00

12. The allowable deviation from the nominal dimension of a part feature ($\pm 0.005''$, $\pm 1/64''$) describes a:

   A. feature.

   B. tolerance.

   C. datum.

   D. position.

13. When calling turret position #7 and the corresponding offset, which of the following words is correct?

   A. T0707

   B. T77

   C. T#7 T#7

   D. 7070 T

14. Select the answer that **best** describes the Jog mode function on a CNC lathe.

   A. Allows programs to be modified

   B. Sends machine to its reference position

   C. Must be selected to run a program

   D. Allows manual control of axes movement

15. In the following image, arrow "A" identifies the _____ axis, and arrow "B" identifies the _____ axis.

   A. *Z*; *X*   C. *X*; *Z*

   B. *X*; *Y*   D. *Y*; *Z*

16. G96 is the code used to initiate:

   A. tool change.   C. constant surface feed.

   B. boring cycle.   D. work plane selection.

17. Which of the following specifications specifies the smoothest surface finish?

   A. 60 µin   C. $\sqrt[12]{}$

   B. 12 psi   D. 25 mm

18. The tool in the following image is a:

   A. counterbore.   C. insert drill.

   B. boring bar.   D. right-hand turning tool.

19. Which sequence of commands will index the turret from position #1 to position #8?

   A. Select Jog mode, type "T01 GOTO T08," Enter, Cycle Start

   B. Select Edit mode, type "T08," End of Block, Enter, Cycle Start

   C. Select Manual Data Input mode, type "T08," End of Block, Enter, Cycle Start

   D. All of the above

   E. None of the above

20. Select the answer that **best** describes the Edit mode function.

   A. Allows programs to be modified
   B. Allows manual control of the axes movement
   C. Provides the screen to input and execute single commands
   D. The mode selected when a program is running

21. Which code corresponds with rapid positioning?

   A. G00
   B. M00
   C. G03
   D. M03

22. What does the symbol ⟋⟋ specify?

   A. Total runout
   B. Symmetry
   C. Straightness
   D. Parallelism

23. A CNC machinist is writing a program to produce a part from plain carbon steel. Which roughing and finishing feed rates are applicable to safely produce this part?

   A. 0.005 IPR roughing; 0.150 IPR finishing
   B. 0.150 IPR roughing; 0.0005 IPR finishing
   C. 10.0 IPR roughing; 5.0 IPR finishing
   D. 0.015 IPR roughing; 0.005 IPR finishing

24. Which code directs the program to end-return to start?

   A. M02
   B. M99
   C. M30
   D. G80

25. Which of the following symbols specifies profile of a surface?

   A. ⚌
   B. ⌓
   C. ⌖
   D. ⌒

26. After running a setup piece for the following part, the diameters (all finished with T02) are 0.004″ over their nominal dimensions. What will correct this before the next part is machined?

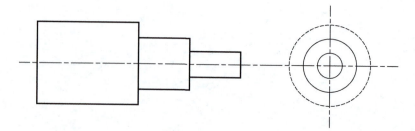

   A. Adjust the tool #2, X wear offset by 0.002″
   B. Adjust the tool #2, Z wear offset by 0.002″
   C. Adjust the tool #2, X wear offset by −0.004″
   D. Adjust the tool #2, Z wear offset by −0.004″

27. Select the answer that **best** describes the offset page.

    A. Displays settings and parameters that are turned on and off frequently

    B. Screen displaying tool length, diameter, and wear offsets

    C. List of machine parameters that are seldom changed

    D. Screen displaying work and machine coordinates as well as distance to go

28. To utilize an M01 command, which mode function would have to be selected?

    A. Block Delete

    B. Select Program

    C. Optional Stop

    D. Single Block

29. Which direction should the wrench in the following image turn in order to tighten the cam of this insert holder?

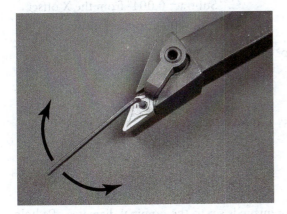

    A. Left

    B. Right

    C. Clockwise

    D. Counterclockwise

30. After rough turning the profile of a part, a program stop is utilized in the program so the metal chips can be cleared. What will restart the machine from this point in the program?

    A. power on

    B. enter

    C. reset

    D. cycle start

31. To start the spindle on a CNC Turning Center in the clockwise direction (viewed from behind the spindle), the machinist would use which of the following codes?

    A. M03

    B. M04

    C. G04

    D. G03

32. A part print specifies a part's finished length to be 1.750″. You decide to leave 1/8″ of material on each piece for machining operations. How many pieces can be cut from a 17.0″ length of material?

    A. 10

    B. 9

    C. 8

    D. 7

33. After running a setup piece for the part in the following image, the inside diameter (machined with tool #4— boring bar) measures 1.8190″. What offset adjustment will bring this dimension to the nominal size?

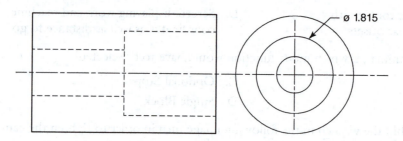

  A. Decrease the X offset by 0.002″   C. Subtract 0.004″ from the X offset

  B. Add 0.002″ to the X offset   D. Add 0.004″ to the X offset

34. Which code will command a program stop?

  A. G00   C. M00

  B. M30   D. M05

35. A print specifies a part's diameter at 2.175″ ± 0.001″. The part measures 2.1775″. How much out of tolerance is this part?

  A. 0.0025″   C. 0.001″

  B. 0.0015″   D. 0.0005″

36. The zero on the bore gage in the following image is set to the nominal diameter of a hole. Based on the measurement in the image, what adjustment to the boring bar is necessary to machine the bore to the nominal size on the next piece?

  A. −0.0002   C. −0.025″

  B. +0.0025   D. −0.0025″

37. Which statement **best** describes a datum?

  A. A rectangular box with a series of compartments for symbols and dimensions that describe a part feature

  C. Shown by a capital letter in a box and signifies a plane from which dimensions are referenced

  B. A numerical value positioned within a box used to describe a theoretical dimension

  D. Compares the relationship of two or more cylindrical features

38. What is the dimension labeled "A" on the following part drawing considered in Geometric Dimensioning and Tolerancing?

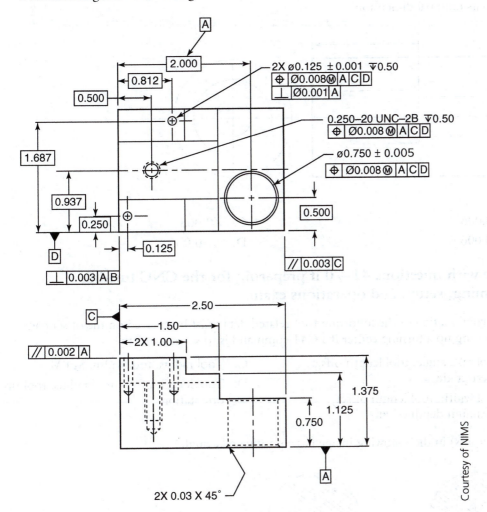

Courtesy of NIMS

A. Parallel
B. Basic

C. Datum
D. Feature control frame

39. Which display screen page are tool length data stored in?

A. Program library
B. Message page

C. Offset page
D. Parameter screen

40. After machining the part in the following image for several shifts, an inspection reveals the depth of the bore is 1.997″. What wear offset adjustment will bring the depth of this bore back to its nominal dimension?

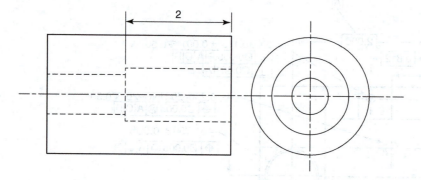

A. X0.006

B. Z0.006

C. Z0.003

D. Z−0.003

Continue with questions 41–70 if preparing for the CNC turning programming, setup, and operations exam.

41. Which combination of the following tool-related data must be entered on the offset page when setting up a turning center if a G41 command is to work correctly?

A. Tool rake angle, tool length offset, insert grade

B. Tool width, tool length offset, maximum depth of cut

C. Tool radius, tool width, SFPM

D. Tool length offset, tool radius, tool tip orientation

42. Which tool(s) in the following image has a positive rake angle?

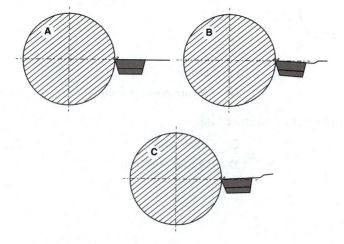

A. A

B. B

C. C

D. Both A and B

43. What is the pitch diameter of a ½-13 UNC 2A thread? *Use The Machinery's Handbook or the Shop Reference for Students and Apprentices as a reference.*

    A. 0.4555

    B. 0.4455

    C. 0.4955

    D. 0.5000

44. Referencing the line of code, what answer correctly describes the "S" word? N35 G96 S300

    A. Surface feet per minute

    B. RPM

    C. Inches per minute

    D. Minimum RPM

45. Which of the following codes will initiate a rough turning cycle?

    A. G70

    B. G71

    C. G80

    D. G81

46. Which of the following answers corresponds to a G70 code?

    A. Can cycle cancel

    B. Absolute programming

    C. Rough turning cycle

    D. Finish turning cycle

47. CNC commands that remain active until canceled or overridden are called:

    A. current.   C. preparatory.

    B. modal.   D. polar.

Refer to the program in Figure 9.12 and the part print in Figure 9.13 to answer questions 48–70.

CNC Turning Program

```
Tool 1-Rough Face/Turn Tool -.031" nose radius
Tool 2-Finish Face/Turn Tool -.015" nose radius

%
O5555(LATHE PART);
N5 G28;
N10 G0 G20 G99;
N15 T0101;
N20 G_____ S1500;
N25 G97 S300 M_____;
N30 G0 Z0.010;
N35 X2.8;
N40 G_____ S350;
N45 G01 X_____ F.010;
N50 G0 Z.150;
N55 X2.550;
N60 G42 Z.100;
N65 G____ P70 Q120 U.025 W.010 D.125 F.012;
N70 G0 X_____;
N75 Z.010;
N80 G01 X.875 Z-.090 F.010;
N85 Z-.500;
N90 X_____ Z_____;
N95 Z-1.725;
N100 G___ X____ Z____ I.125 K0.0;
N105 ____ X2.160;
N110 G03 X2.410 Z-1.975 I_____ K_____;
N115 G01 Z-2.750;
N120 X2.6;
N125 G97;
N130 G0 G40 Z.500;
N135 M01;
N140 G28;
N145 T0202;
N150 G50 S2200;
N155 G97 S430;
N160 G0 Z0.0;
N165 X1.5;
N170 G96 S450;
N175 G42 X1.0;
N180 G01 X_____ F.005;
N185 X.700;
N190 G_____ P80 Q_____ F.005;
N195 G97;
N200 M05;
N205 G28;
N210 _____;
%
```

48. On line N20, what code will set the maximum RPM?

    A. G20
    B. G28
    C. G50
    D. G70

49. On line N10, what does G99 establish?

    A. Absolute programming
    B. Incremental programming
    C. Reference position
    D. Per revolution feed

50. On line N40, which code will activate constant surface speed?

    A. G96
    B. G92
    C. G90
    D. G41

51. On line N25, what code will start the spindle?

    A. M03
    B. G03
    C. G04
    D. M01

52. On line N45 what X coordinate is the **best** choice for completing the facing operation?

    A. X0.0
    B. X$-$0.005
    C. X$-$0.040
    D. X.010

53. What will the G42 on line N60 initiate?

    A. Constant surface speed
    B. Cutter compensation right
    C. Cutter compensation left
    D. Inch units

54. On line N65, what code will initiate a rough turning cycle?

    A. G21
    B. G41
    C. G61
    D. G71

55. On line N70, what does the "P" value specify?

    A. The number of passes in the cycle
    B. The depth of cut
    C. The finish allowance
    D. The line number to begin the roughing cycle

56. On line N65, what does W.010 indicate?

    A. *X*-axis stock allowance left for finishing
    B. *Z*-axis stock allowance left for finishing
    C. Depth of cut
    D. Feed rate

57. On line N70, what is the *X* value (position 4)?

    A. 0.700
    B. 0.875
    C. 0.785
    D. 0.830

58. On line N100, what is the applicable command to cut the arc? (points 8–9)

    A. G01
    B. G02
    C. G03
    D. G04

59. On line N100, what are the X and Z coordinates?

    A. X1.600, Z1.850
    B. X1.500, Z$-$1.850
    C. X1.350, Z$-$1.600
    D. X1.600, Z$-$1.850

60. On line N105, what code will command the program to cut from point 9 to point 10?

    A. M01   C. M00
    B. G00   D. G01

61. On line N110, what are the I and K values for cutting the arc? (Points 10–11)

    A. I0.0, K0.0   C. I.250, K0.0
    B. I−0.125, K0.0   D. I0.0, K−0.125

62. On line N90, what are the X and Z coordinates? (Position 7)

    A. X1.350, Z1.100   C. X1.350, Z−1.100
    B. X.675, Z−1.100   D. X.675, Z1.100

63. What function does the G40 code on line N130 initiate?

    A. Cutter compensation cancel   C. Spindle stop
    B. Constant surface speed cancel   D. Dwell

64. What function does the M01 code on line N135 initiate?

    A. Program stop   C. Macro call
    B. Can cycle cancel   D. Optional stop

65. On line N180, which of the following X coordinates is the **best** option for the finish facing operation?

    A. X−.005   C. X.020
    B. X−0.020   D. X0.0

66. On line N190, what code will initiate a finish turning cycle?

    A. G90   C. G70
    B. G80   D. G76

67. On line N190, what is the best choice for the "Q" word?

    A. 180   C. 120
    B. 170   D. 102

68. On line N205, what is the function of the G28 code?

    A. Return to reference position   C. Coolant stop
    B. Spindle stop   D. Program stop

69. On line N210, what code will end the program and return to the start?

    A. M03   C. M30
    B. G03   D. M00

70. On line N110, instead of using the I and K words to program the arc, an R word can be used. Which of the following lines is correct for cutting this arc using R? (Points 10–11)

    A. G02 X 2.410 Z−1.975 R.125   C. G03 X 2.410 Z−1.975 R.125
    B. G02 X 2.410 Z−1.850 R.250   D. G03 X 2.410 Z−1.950 R.125

9.2 ANSWER KEY

1.	B	26.	C	51.	A
2.	C	27.	B	52.	C
3.	A	28.	C	53.	B
4.	E	29.	C	54.	D
5.	D	30.	D	55.	D
6.	B	31.	A	56.	B
7.	B	32.	C	57.	A
8.	D	33.	C	58.	B
9.	D	34.	C	59.	D
10.	A	35.	B	60.	D
11.	A	36.	D	61.	D
12.	B	37.	C	62.	C
13.	A	38.	B	63.	A
14.	D	39.	C	64.	D
15.	C	40.	D	65.	B
16.	C	41.	D	66.	C
17.	C	42.	B	67.	C
18.	D	43.	B	68.	A
19.	C	44.	A	69.	C
20.	A	45.	B	70.	C
21.	A	46.	D		
22.	A	47.	B		
23.	D	48.	C		
24.	C	49.	D		
25.	B	50.	A		

9.3 PRACTICE THEORY EXAM EXPLANATIONS

1. Typically done as part of the start-up procedure, which of the following is the process of positioning the CNC turning center's axes at a fixed reference position to allow the MCU to accurately establish/track the machine's position?

    A. Jogging   C. Calibrating
    B. Homing   D. Positioning

    Answer B is correct. Homing is the act of positioning the machine's axes at their reference position, also called machine zero position. Once at this position, information is relayed back to the MCU and the coordinates are verified. This is typically accomplished automatically by pressing a button(s) or key(s) on the machine's control panel.

2. To maintain proper coolant levels and concentrations, and to ensure a machine tool has sufficient lubrication, a machinist should check the machine's fluids:

    A. at the beginning of each week.   C. every shift.
    B. at the end of each week.   D. every other week.

    Answer C is correct. Every machine tool manufacturer has its own recommendations for maintenance intervals and procedures. To avoid unnecessary wear to the machining center and to maintain optimal cutting conditions, the coolant and lubrication system should be checked at the start of every shift.

3. Which code identifies linear interpolation (feed in a straight line)?

    A. G01   C. M01
    B. G00   D. G20

    Answer A is correct. G01 is the code for linear interpolation or straight-line movement controlled by a feed rate.

4. Your employer has appointed you to work with the safety committee to develop a department safety manual. Which safety rule listed here will help facilitate the safest work environment for the employees in this department?

    A. Store all tools in their appropriate location.   D. Disconnect power sources when servicing a machine.
    B. Do not bypass door interlocks.   E. Follow all safety rules at all times.
    C. Wear safety glasses at all times.

    Answer E is correct. To provide the highest level of safety and protection from hazards, all safety rules should be followed at all times.

5. Referring to Geometric Dimensioning and Tolerancing, which of the following **best** describes a basic dimension?

    A. Title block tolerances are not applied to a basic dimension.   C. It has a box around the dimension.
    B. A basic dimension is a numerical value used to describe a theoretical size, profile, orientation, or datum target.   D. All of the above are correct.
                                                                              E. B and C only are correct.

    Answer D is correct. A basic dimension is represented by a boxed dimension and is a theoretical precise value used to designate the exact size or location of a geometric feature.

6. Which statement **best** describes geometry offsets?

    A. Adjusted small amounts to compensate for tool wear and/or thermal variation

    B. Provides tool-specific data to the control, such as the length, nose radius, and tip position

    C. Sets the X and Z origin of the part

    D. Is accessed using G54

    Answer B is correct. The geometry offset page on a typical CNC control stores data pertaining to the tool's length, nose radius, and tip position.

7. Which of the following tools is a right-hand turning tool and will work correctly with the M03 command when installed in a standard slant bed CNC lathe?

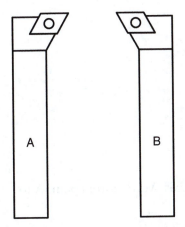

    A. Tool A fits this description.

    B. Tool B fits this description.

    C. Both tools fit this description.

    D. Neither tool fits this description.

    Answer B is correct. When viewing the spindle from behind, the M03 command will start the spindle's rotation in the clockwise direction. Considering the cutting will take place on the opposite side of the workpiece when compared to a conventional lathe, tool B is a right-hand cutting tool and would be installed with the insert facing down.

8. Select the mode switch position used to return a machine's axes to their home position.

    A. Auto

    B. MDI

    C. Edit

    D. Zero return

    Answer D is correct. Zero return will position the machine's axes at their reference position or machine zero position.

9. Which of the methods listed here is recommended when adding fluid to a coolant tank?

    A. Add water until the desired level is reached.

    B. Mix a 50:50 soluble oil to water mixture and add it to the tank.

    C. Mix a 20:80 soluble oil to water mixture and add it to the tank.

    D. Check the coolant ratio with a refractometer. Add soluble oil and/or water as necessary to obtain the desired concentration.

    Answer D is correct. To provide the maximum cooling and lubricity, water-soluble cutting fluids must be properly maintained. The concentration level should be checked with a refractometer and adjusted according to the manufacturer's recommendations.

10. A machinist needs to jog the tool from point A to point B in the following image. Which axis and in what direction should the tool be moved?

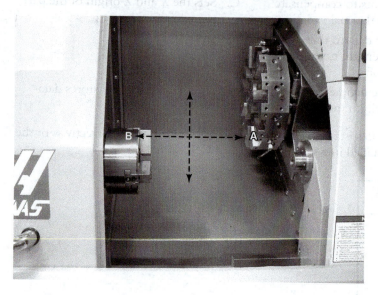

A. *Z*; –

B. *X*; –

C. *Y*; –

D. *X*; +

Answer A is correct. The arrow is positioned along the *Z*-axis. Moving from point A to point B will move the tool in the negative direction.

11. Which code directs a CNC Turning Center to move the machine's axes to their home positions?

A. G28

B. G20

C. G90

D. G00

Answer A is correct. Calling a G28 in a CNC program will rapid the machine's axes to their home position.

12. The allowable deviation from the nominal dimension of a part feature ($\pm 0.005''$, $\pm 1/64''$) describes a:

A. feature.

B. tolerance.

C. datum.

D. position.

Answer B is correct. Tolerance is the amount a geometric feature is allowed to vary from its specified nominal size.

13. When calling turret position #7 and the corresponding offset, which of the following words is correct?

A. T0707

B. T77

C. T#7 T#7

D. 7070 T

Answer A is correct. The format for a tool change on most turning centers' controls is T followed by the tool number (ex. 07 or 12), followed by the offset location (ex. 07 or 12).

14. Select the answer that **best** describes the Jog mode function on a CNC lathe.

    A. Allows programs to be modified
    B. Sends machine to its reference position
    C. Must be selected to run a program
    D. Allows manual control of axes movement

    Answer D is correct. Manual control of the machine's axes through the use of buttons or a rotary hand wheel is accomplished in the Jog mode.

15. In the following image, arrow "A" identifies the _____ axis, and arrow "B" identifies the _____ axis.

    A. Z; X
    B. X; Y
    C. X; Z
    D. Y; Z

    Answer C is correct. Arrow "A" is positioned along the X-axis, which is perpendicular to the axis of the spindle. Arrow "B" is positioned along the Z-axis, which is parallel to the axis of the spindle.

16. G96 is the code used to initiate:

    A. tool change.
    B. boring cycle.
    C. constant surface feed.
    D. work plane selection.

    Answer C is correct. Calling a G96 in a CNC program will initiate constant surface speed. Constant surface speed will adjust the spindle's RPM to maintain the cutting speed in surface feet per minute.

17. Which of the following specifications specifies the smoothest surface finish?

    A. 60 μin
    B. 12 psi
    C. $\sqrt[12]{}$
    D. 25 mm

    Answer C is correct. The smaller the value is in microinches, the smoother the surface finish will be. The symbol $\sqrt[12]{}$ is less than 60 μin and is therefore the smoothest surface.

18. The tool in the following image is a:

A. counterbore.

B. boring bar.

C. insert drill.

D. right-hand turning tool.

Answer D is correct. This tool is a right-hand turning tool. When installed in a slant bed CNC lathe, the insert will face down. This orientation allows the tool to cut with the M03 command and feed toward the workholding device.

19. Which sequence of commands will index the turret from position #1 to position #8?

A. Select Jog mode, type "T01 GOTO T08," Enter, Cycle Start

B. Select Edit mode, type "T08," End of Block, Enter, Cycle Start

C. Select Manual Data Input mode, type "T08," End of Block, Enter, Cycle Start

D. All of the above

E. None of the above

Answer C is correct. Of the choices listed, using the Manual Data Input mode and inputting T08 and then pressing the Cycle Start button will index the turret to tool number 8.

20. Select the answer that **best** describes the Edit mode function.

A. Allows programs to be modified

B. Allows manual control of the axes movement

C. Provides the screen to input and execute single commands

D. The mode selected when a program is running

Answer A is correct. The Edit mode on a CNC control is where new programs can be entered in to the control, existing programs can be modified, and programs can be installed prior to use.

21. Which code corresponds with rapid positioning?

A. G00

B. M00

C. G03

D. M03

Answer A is correct. Rapid traverse or rapid positioning occurs when a machining center's axes are moved quickly to a specified position. This is accomplished in a program with a G00 code.

22. What does the symbol ⟋⟋ specify?

    A. Total runout
    C. Straightness
    B. Symmetry
    D. Parallelism

    Answer A is correct. The Geometric Dimensioning and Tolerancing symbol ⟋⟋ indicates a specified total runout tolerance over a cylindrical surface.

23. A CNC machinist is writing a program to produce a part from plain carbon steel. Which roughing and finishing feed rates are applicable to safely produce this part?

    A. 0.005 IPR roughing; 0.150 IPR finishing
    C. 10.0 IPR roughing; 5.0 IPR finishing
    B. 0.150 IPR roughing; 0.0005 IPR finishing
    D. 0.015 IPR roughing; 0.005 IPR finishing

    Answer D is correct. For roughing cuts (<.250″) where the maximum material removal rate is the primary consideration, a feed rate of 0.015 IPR is applicable. For finishing cuts (<.050″) where the surface finish is the primary consideration, a feed rate of 0.005 IPR is applicable.

24. Which code directs the program to end-return to start?

    A. M02
    C. M30
    B. M99
    D. G80

    Answer C is correct. M30 is the M code that will end the program and reset the program back from the beginning.

25. Which of the following symbols specifies profile of a surface?

    A. ⹀
    C. ⌀
    B. ⌒
    D. ⌒

    Answer B is correct. The Geometric Dimensioning and Tolerancing symbol ⌒ indicates all cross sections of a surface when compared to a specified datum must be within the tolerance limits.

26. After running a setup piece for the following part, the diameters (all finished with T02) are .004″ over their nominal dimensions. What will correct this before the next part is machined?

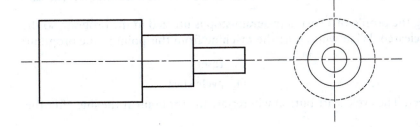

    A. Adjust the tool #2, X wear offset by .002″
    C. Adjust the tool #2, X wear offset by −0.004″
    B. Adjust the tool #2, Z wear offset by .002″
    D. Adjust the tool #2, Z wear offset by −0.004″

    Answer C is correct. The dimensions measure .004″ over their nominal dimensions. Adjusting the X wear offset for this tool in the wear offset page will command this tool to cut closer to the programmed size by .002″ per side, or .004″ on the diameter.

27. Select the answer that best describes the offset page.

    A. Displays settings and parameters that are turned on and off frequently

    B. Screen displaying tool length, diameter, and wear offsets

    C. List of machine parameters that are seldom changed

    D. Screen displaying work and machine coordinates as well as distance to go

    Answer B is correct. The offset page on a typical CNC control consists of three screens to store offset-related data. The three pages are the geometry offsets, wear offsets, and workpiece offset.

28. To utilize an M01 command, which mode function would have to be selected?

    A. Block Delete

    B. Select Program

    C. Optional Stop

    D. Single Block

    Answer C is correct. The M01 code will stop the program when the optional stop button on the control is enabled.

29. Which direction should the wrench in the following image turn in order to tighten the cam of this insert holder?

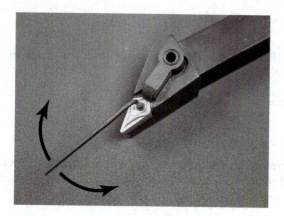

    A. Left

    B. Right

    C. Clockwise

    D. Counterclockwise

    Answer C is correct. Turning the wrench in the clockwise direction will tighten the cam.

30. After rough turning the profile of a part, a program stop is utilized in the program so the metal chips can be cleared. What will restart the machine from this point in the program?

    A. power on

    B. enter

    C. reset

    D. cycle start

    Answer D is correct. The cycle start button will restart the program at the line after the M00 code.

31. To start the spindle on a CNC Turning Center in the clockwise direction (viewed from behind the spindle), the machinist would use which of the following codes?

    A. M03   C. G04

    B. M04   D. G03

    Answer A is correct. M03 is the code to start the spindle in the clockwise direction.

32. A part print specifies a part's finished length to be 1.750″. You decide to leave 1/8″ of material on each piece for machining operations. How many pieces can be cut from a 17.0″ length of material?

    A. 10   C. 8

    B. 9   D. 7

    Answer C is correct. The thickness of the blade on the stock saw must also be taken in to consideration. A typical bandsaw blade is approximately .040″ thick. $17.0 \div (1.75 + .125 + .040) = 8.88$, or eight full pieces.

33. After running a setup piece for the part in the following image, the inside diameter (machined with tool #4– boring bar) measures 1.8190″. What offset adjustment will bring this dimension to the nominal size?

    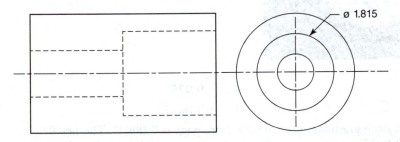

    A. Decrease the X offset by .002″   C. Subtract .004″ from the X offset

    B. Add .002″ to the X offset   D. Add .004″ to the X offset

    Answer C is correct. The dimension measures .004″ over the nominal dimension. Adjusting the X wear offset for this tool in the wear offset page will command this tool to cut closer to the programmed size by .002″ per side, or 0.004″ on the diameter.

34. Which code will command a program stop?

    A. G00   C. M00

    B. M30   D. M05

    Answer C is correct. The M00 code will stop the program.

35. A print specifies a part's diameter at 2.175″ ± 0.001″. The part measures 2.1775″. How much out of tolerance is this part?

   A. 0.0025″

   B. 0.0015″

   C. 0.001″

   D. 0.0005″

   Answer B is correct. The upper limit of this dimension would be determined by: 2.175″ + 0.001″ = 2.176″. To determine how much part is over the upper limit, 2.1775″ − 2.176″ = 0.0015″.

36. The zero on the bore gage in the following image is set to the nominal diameter of a hole. Based on the measurement in the image, what adjustment to the boring bar is necessary to machine the bore to the nominal size on the next piece?

   A. −0.0002

   B. +0.0025

   C. −0.025″

   D. −0.0025″

   Answer D is correct. Each graduation on the dial bore gage is 0.0001″. The needle is 25 graduations or 0.0025″ past the zero. The boring bar would have to be adjusted −0.0025″ to cut to the nominal size.

37. Which statement **best** describes a datum?

   A. A rectangular box with a series of compartments for symbols and dimensions that describe a part feature

   B. A numerical value positioned within a box used to describe a theoretical dimension

   C. Shown by a capital letter in a box and signifies a plane from which dimensions are referenced

   D. Compares the relationship of two or more cylindrical features

   Answer C is correct. A datum is a point, axis, or plane from which the locations of the part's geometric features are referenced. A datum is specified by a capital letter inside a square box.

38. What is the dimension labeled "A" on the following part drawing considered in Geometric Dimensioning and Tolerancing?

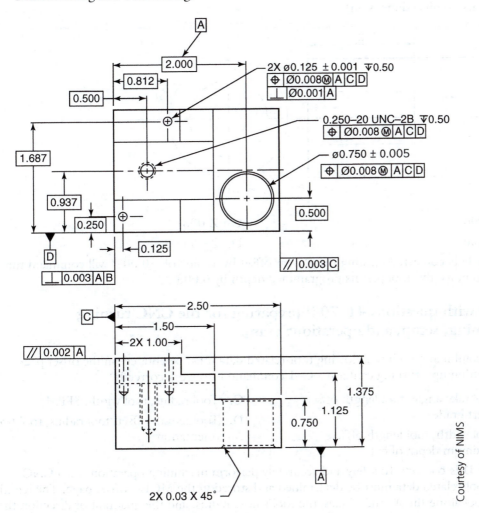

A. Parallel

B. Basic

C. Datum

D. Feature control frame

Answer B is correct. A basic dimension is represented by a boxed dimension and is a theoretical precise value used to designate the exact size or location of a geometric feature.

39. Which display screen page are tool length data stored in?

A. Program library

B. Message page

C. Offset page

D. Parameter screen

Answer C is correct. Tool length values are stored in the geometry offset page.

40. After machining the part in the following image for several shifts, an inspection reveals the depth of the bore is 1.997". What wear offset adjustment will bring the depth of this bore back to its nominal dimension?

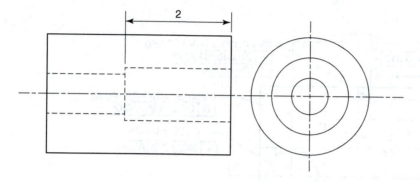

A. X.006

B. Z.006

C. Z.003

D. Z−0.003

Answer D is correct. Adjusting the Z wear offset by a value of −0.003" will command the MCU to move this tool past its programmed depth by 0.003".

Continue with questions 41–70 if preparing for the CNC turning programming, setup, and operations exam.

41. Which combination of the following tool-related data must be entered on the offset page when setting up a turning center if a G41 command is to work correctly?

A. Tool rake angle, tool length offset, insert grade

B. Tool width, tool length offset, maximum depth of cut

C. Tool radius, tool width, SFPM

D. Tool length offset, tool radius, tool tip orientation

Answer D is correct. To safely and accurately perform machining operations on a CNC lathe, tool-related data must be determined and stored in the MCU's offset page. The length of the tool along the X- and Z-axes, the tool's nose radius, and the quadrant or direction the tool's tip is positioned are all required if cutter compensation is to function properly.

42. Which tool(s) in the following image has a positive rake angle?

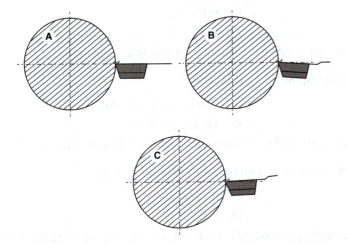

A. A

B. B

C. C

D. Both A and B

Answer B is correct. The back rake angle or the angle on the top of the cutting tool can have neutral, positive, or negative positioning. A positive rake tool penetrates the work with the tip of the tool first.

43. What is the pitch diameter of a ½-13 UNC 2A thread? Use *The Machinery's Handbook* or the *Shop Reference for Students and Apprentices* as a reference.

A. 0.4555

B. 0.4455

C. 0.4955

D. 0.5000

Answer B is correct. In this example the chart is specifying a pitch diameter range of 0.4435″ to 0.4485″. Answer B, 0.4455, is within this range.

44. Referencing the line of code, what answer correctly describes the "S" word? N35 G96 S300

A. Surface feet per minute

B. RPM

C. Inches per minute

D. Minimum RPM

Answer A is correct. When utilizing a G96 (constant surface speed) command, the S word provides the MCU with the surface feet per minute value to regulate the spindle RPM.

45. Which of the following codes will initiate a rough turning cycle?

    A. G70   C. G80

    B. G71   D. G81

Answer B is correct. G71 is the code used to activate the roughing canned cycle.

46. Which of the following answers corresponds to a G70 code?

    A. Can cycle cancel   C. Rough turning cycle

    B. Absolute programming   D. Finish turning cycle

Answer D is correct. G70 is the code used to activate the finish turning canned cycle.

47. CNC commands that remain active until canceled or overridden are called:

    A. current.   C. preparatory.

    B. modal.   D. polar.

Answer B is correct. Some preparatory commands remain on or active upon being read by the MCU. These commands are called modal and will stay on until canceled or changed by another, conflicting preparatory command.

Refer to the program in Figure 9.12 and the part print in Figure 9.13 to answer questions 48–70.

CNC Turning Program

```
Tool 1-Rough Face/Turn Tool -.031" nose radius
Tool 2-Finish Face/Turn Tool -.015" nose radius

%
O5555(LATHE PART);
N5 G28;
N10 G0 G20 G99;
N15 T0101;
N20 G_____ S1500;
N25 G97 S300 M_____;
N30 G0 Z0.010;
N35 X2.8;
N40 G_____ S350;
N45 G01 X_____ F.010;
N50 G0 Z.150;
N55 X2.550;
N60 G42 Z.100;
N65 G_____ P70 Q120 U.025 W.010 D.125 F.012;
N70 G0 X_____;
N75 Z.010;
N80 G01 X.875 Z-.090 F.010;
N85 Z-.500;
N90 X_____ Z_____;
N95 Z-1.725;
N100 G___ X____ Z____ I.125 K0.0;
N105 ____ X2.160;
N110 G03 X2.410 Z-1.975 I____ K____;
N115 G01 Z-2.750;
N120 X2.6;
N125 G97;
```

```
N130 G0 G40 Z.500;
N135 M01;
N140 G28;
N145 T0202;
N150 G50 S2200;
N155 G97 S430;
N160 G0 Z0.0;
N165 X1.5;
N170 G96 S450;
N175 G42 X1.0;
N180 G01 X_____ F.005;
N185 X.700;
N190 G_____ P80 Q_____ F.005;
N195 G97;
N200 M05;
N205 G28;
N210 _____;
%
```

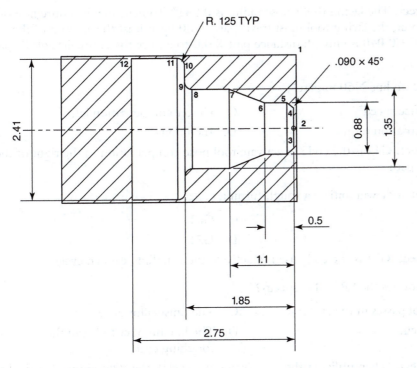

48. On line N20, what code will set the maximum RPM?

    A. G20   C. G50
    B. G28   D. G70

    Answer C is correct. Utilizing a G50 code with an S word value will establish and set the maximum RPM.

49. On line N10, what does G99 establish?

    A. Absolute programming   C. Reference position
    B. Incremental programming   D. Per revolution feed

    Answer D is correct. G99 is the code to initiate feed in inches per revolution.

50. On line N40, which code will activate constant surface speed?

    A. G96   C. G90

    B. G92   D. G41

Answer A is correct. G96 is the code to initiate constant surface speed, which will adjust the spindle's RPM to a specified surface feet per minute value.

51. On line N25, what code will start the spindle?

    A. M03   C. G04

    B. G03   D. M01

Answer A is correct. M03 is the code to start the spindle in the clockwise direction.

52. On line N45 what X coordinate is the best choice for completing the facing operation?

    A. X0.0   C. X−0.040

    B. X−0.005   D. X.010

Answer C is correct. The facing tool's nose radius is 0.031″. To machine the entire face of the part to the center, the facing tool must travel past X 0.0 by at least the amount of the tool nose radius. X −0.040 is enough distance past X 0.0 to ensure the entire face of the part is machined.

53. What will the G42 on line N60 initiate?

    A. Constant surface speed   C. Cutter compensation left

    B. Cutter compensation right   D. Inch units

Answer B is correct. G42 is the code to activate tool nose compensation to the right of the programmed tool path.

54. On line N65, what code will initiate a rough turning cycle?

    A. G21   C. G61

    B. G41   D. G71

Answer D is correct. G71 is the code used to activate the roughing canned cycle.

55. On line N70, what does the "P" value specify?

    A. The number of passes in the cycle   C. The finish allowance

    B. The depth of cut   D. The line number to begin the roughing cycle

Answer D is correct. When utilizing the G71 (roughing cycle), the P word specifies the line number where the profile's tool path starts.

56. On line N65, what does W.010 indicate?

    A. *X*-axis stock allowance left for finishing   C. Depth of cut

    B. *Z*-axis stock allowance left for finishing   D. Feed rate

Answer B is correct. When utilizing the G71 (roughing cycle), the W word specifies the material allowance to be left on all of the faces and shoulders for the finish operation.

57. On line N70, what is the *X* value (position 4)?

    A. 0.700   C. 0.785

    B. 0.875   D. 0.830

Answer A is correct. The diametral *X*-axis value for position 4 is 0.700 ($0.880 − 2 \times 0.090$).

58. On line N100, what is the applicable command to cut the arc? (points 8–9)

   A. G01   C. G03
   B. G02   D. G04

   Answer B is correct. The arc formed by points 8–9 is an inside arc and will require the turning tool to cut in the clockwise direction. G02 is the code for circular interpolation clockwise.

59. On line N100, what are the X and Z coordinates?

   A. X1.600, Z1.850   C. X1.350, Z−1.600
   B. X1.500, Z−1.850   D. X1.600, Z−1.850

   Answer D is correct. After the G02 command is initiated, the X and Z coordinates represent the endpoint of the arc. In this case they are X1.600, Z−1.850.

60. On line N105, what code will command the program to cut from point 9 to point 10?

   A. M01   C. M00
   B. G00   D. G01

   Answer D is correct. The part feature from point 9 to 10 is a straight-line move on the X-axis. G01 is the code for linear interpolation or straight-line movement controlled by a feed rate.

61. On line N110, what are the I and K values for cutting the arc? (Points 10–11)

   A. I0.0, K0.0   C. I.250, K0.0
   B. I−0.125, K0.0   D. I0.0, K−0.125

   Answer D is correct. The I and K values correspond to the X- and Z-axes and are the incremental distance from the start point to the center point of the arc. The radius is 0.125″, and the center is positioned at the same X coordinate as the start point, which results in I0.0. The Z coordinate of the center point is 0.125″ in the negative direction from the Z coordinate of the start point, resulting in K−0.125″.

62. On line N90, what are the X and Z coordinates? (Position 7)

   A. X1.350, Z1.100   C. X1.350, Z−1.100
   B. X.675, Z−1.100   D. X.675, Z1.100

   Answer C is correct. The coordinates of point 7 are X1.350, Z−1.100.

63. What function does the G40 code on line N130 initiate?

   A. Cutter compensation cancel   C. Spindle stop
   B. Constant surface speed cancel   D. Dwell

   Answer A is correct. G40 is the code to cancel cutter (tool nose) compensation.

64. What function does the M01 code on line N135 initiate?

   A. Program stop   C. Macro call
   B. Can cycle cancel   D. Optional stop

   Answer D is correct. The M01 code will stop the program when the optional stop button on the control is enabled.

65. On line N180, which of the following X coordinates is the best option for the finish facing operation?

    A. X−0.005

    B. X−0.020

    C. X.020

    D. X0.0

    Answer B is correct. The finish facing tool's nose radius is 0.015″. In order to machine the entire face of the part to the center, the facing tool must travel past X0.0 by at least the amount of the tool nose radius. X−0.020 is enough distance past X0.0 to ensure the entire face of the part is machined.

66. On line N190, what code will initiate a finish turning cycle?

    A. G90

    B. G80

    C. G70

    D. G76

    Answer C is correct. G70 is the code used to activate the finish turning canned cycle.

67. On line N190, what is the best choice for the "Q" word?

    A. 180

    B. 170

    C. 120

    D. 102

    Answer C is correct. When utilizing the G70 (finishing cycle), the Q word specifies the line number where the profile's tool path ends.

68. On line N205, what is the function of the G28 code?

    A. Return to reference position

    B. Spindle stop

    C. Coolant stop

    D. Program stop

    Answer A is correct. Calling a G28 in a CNC program will rapid the machine's axes to their home position.

69. On line N210, what code will end the program and return to the start?

    A. M03

    B. G03

    C. M30

    D. M00

    Answer C is correct. M30 is the code to end a program and return to the start.

70. On line N110, instead of using the I and K words to program the arc, an R word can be used. Which of the following lines is correct for cutting this arc using R? (Points 10–11)

    A. G02 X 2.410 Z−1.975 R.125

    B. G02 X 2.410 Z−1.850 R.250

    C. G03 X 2.410 Z−1.975 R.125

    D. G03 X 2.410 Z−1.950 R.125

    Answer C is correct. An arc can be programmed by specifying the arc's radius (R) rather than the arc's center point coordinates (I, K). The arc formed by points 10–11 is an outside radius, which would be cut in the counterclockwise direction. G03 is the code for circular interpolation counterclockwise.

Appendices

ANSWER SHEET

1. _____
2. _____
3. _____
4. _____
5. _____
6. _____
7. _____
8. _____
9. _____
10. _____
11. _____
12. _____
13. _____
14. _____
15. _____
16. _____
17. _____
18. _____
19. _____
20. _____
21. _____
22. _____
23. _____
24. _____
25. _____

26. _____
27. _____
28. _____
29. _____
30. _____
31. _____
32. _____
33. _____
34. _____
35. _____
36. _____
37. _____
38. _____
39. _____
40. _____
41. _____
42. _____
43. _____
44. _____
45. _____
46. _____
47. _____
48. _____
49. _____
50. _____

Chapters 8 & 9

51. _____
52. _____
53. _____
54. _____
55. _____
56. _____
57. _____
58. _____
59. _____
60. _____
61. _____
62. _____
63. _____
64. _____
65. _____
66. _____
67. _____
68. _____
69. _____
70. _____

ANSWER SHEET

1. _____	26. _____	**Chapters 8 & 9**
2. _____	27. _____	51. _____
3. _____	28. _____	52. _____
4. _____	29. _____	53. _____
5. _____	30. _____	54. _____
6. _____	31. _____	55. _____
7. _____	32. _____	56. _____
8. _____	33. _____	57. _____
9. _____	34. _____	58. _____
10. _____	35. _____	59. _____
11. _____	36. _____	60. _____
12. _____	37. _____	61. _____
13. _____	38. _____	62. _____
14. _____	39. _____	63. _____
15. _____	40. _____	64. _____
16. _____	41. _____	65. _____
17. _____	42. _____	66. _____
18. _____	43. _____	67. _____
19. _____	44. _____	68. _____
20. _____	45. _____	69. _____
21. _____	46. _____	70. _____
22. _____	47. _____	
23. _____	48. _____	
24. _____	49. _____	
25. _____	50. _____	

ANSWER SHEET

1. _____	26. _____	**Chapters 8 & 9**
2. _____	27. _____	51. _____
3. _____	28. _____	52. _____
4. _____	29. _____	53. _____
5. _____	30. _____	54. _____
6. _____	31. _____	55. _____
7. _____	32. _____	56. _____
8. _____	33. _____	57. _____
9. _____	34. _____	58. _____
10. _____	35. _____	59. _____
11. _____	36. _____	60. _____
12. _____	37. _____	61. _____
13. _____	38. _____	62. _____
14. _____	39. _____	63. _____
15. _____	40. _____	64. _____
16. _____	41. _____	65. _____
17. _____	42. _____	66. _____
18. _____	43. _____	67. _____
19. _____	44. _____	68. _____
20. _____	45. _____	69. _____
21. _____	46. _____	70. _____
22. _____	47. _____	
23. _____	48. _____	
24. _____	49. _____	
25. _____	50. _____	

ANSWER SHEET

1. _____
2. _____
3. _____
4. _____
5. _____
6. _____
7. _____
8. _____
9. _____
10. _____
11. _____
12. _____
13. _____
14. _____
15. _____
16. _____
17. _____
18. _____
19. _____
20. _____
21. _____
22. _____
23. _____
24. _____
25. _____

26. _____
27. _____
28. _____
29. _____
30. _____
31. _____
32. _____
33. _____
34. _____
35. _____
36. _____
37. _____
38. _____
39. _____
40. _____
41. _____
42. _____
43. _____
44. _____
45. _____
46. _____
47. _____
48. _____
49. _____
50. _____

Chapters 8 & 9

51. _____
52. _____
53. _____
54. _____
55. _____
56. _____
57. _____
58. _____
59. _____
60. _____
61. _____
62. _____
63. _____
64. _____
65. _____
66. _____
67. _____
68. _____
69. _____
70. _____

ANSWER SHEET

1. _____	26. _____	**Chapters 8 & 9**
2. _____	27. _____	51. _____
3. _____	28. _____	52. _____
4. _____	29. _____	53. _____
5. _____	30. _____	54. _____
6. _____	31. _____	55. _____
7. _____	32. _____	56. _____
8. _____	33. _____	57. _____
9. _____	34. _____	58. _____
10. _____	35. _____	59. _____
11. _____	36. _____	60. _____
12. _____	37. _____	61. _____
13. _____	38. _____	62. _____
14. _____	39. _____	63. _____
15. _____	40. _____	64. _____
16. _____	41. _____	65. _____
17. _____	42. _____	66. _____
18. _____	43. _____	67. _____
19. _____	44. _____	68. _____
20. _____	45. _____	69. _____
21. _____	46. _____	70. _____
22. _____	47. _____	
23. _____	48. _____	
24. _____	49. _____	
25. _____	50. _____	

ANSWER SHEET

1. _____
2. _____
3. _____
4. _____
5. _____
6. _____
7. _____
8. _____
9. _____
10. _____
11. _____
12. _____
13. _____
14. _____
15. _____
16. _____
17. _____
18. _____
19. _____
20. _____
21. _____
22. _____
23. _____
24. _____
25. _____

26. _____
27. _____
28. _____
29. _____
30. _____
31. _____
32. _____
33. _____
34. _____
35. _____
36. _____
37. _____
38. _____
39. _____
40. _____
41. _____
42. _____
43. _____
44. _____
45. _____
46. _____
47. _____
48. _____
49. _____
50. _____

Chapters 8 & 9

51. _____
52. _____
53. _____
54. _____
55. _____
56. _____
57. _____
58. _____
59. _____
60. _____
61. _____
62. _____
63. _____
64. _____
65. _____
66. _____
67. _____
68. _____
69. _____
70. _____

ANSWER SHEET

1. _____

2. _____

3. _____

4. _____

5. _____

6. _____

7. _____

8. _____

9. _____

10. _____

11. _____

12. _____

13. _____

14. _____

15. _____

16. _____

17. _____

18. _____

19. _____

20. _____

21. _____

22. _____

23. _____

24. _____

25. _____

26. _____

27. _____

28. _____

29. _____

30. _____

31. _____

32. _____

33. _____

34. _____

35. _____

36. _____

37. _____

38. _____

39. _____

40. _____

41. _____

42. _____

43. _____

44. _____

45. _____

46. _____

47. _____

48. _____

49. _____

50. _____

Chapters 8 & 9

51. _____

52. _____

53. _____

54. _____

55. _____

56. _____

57. _____

58. _____

59. _____

60. _____

61. _____

62. _____

63. _____

64. _____

65. _____

66. _____

67. _____

68. _____

69. _____

70. _____

ANSWER SHEET

1. _____
2. _____
3. _____
4. _____
5. _____
6. _____
7. _____
8. _____
9. _____
10. _____
11. _____
12. _____
13. _____
14. _____
15. _____
16. _____
17. _____
18. _____
19. _____
20. _____
21. _____
22. _____
23. _____
24. _____
25. _____

26. _____
27. _____
28. _____
29. _____
30. _____
31. _____
32. _____
33. _____
34. _____
35. _____
36. _____
37. _____
38. _____
39. _____
40. _____
41. _____
42. _____
43. _____
44. _____
45. _____
46. _____
47. _____
48. _____
49. _____
50. _____

Chapters 8 & 9

51. _____
52. _____
53. _____
54. _____
55. _____
56. _____
57. _____
58. _____
59. _____
60. _____
61. _____
62. _____
63. _____
64. _____
65. _____
66. _____
67. _____
68. _____
69. _____
70. _____

Glossary

Abrasive A natural or synthetic particle or grain when bonded to others can be used for metal removal operations such as polishing and grinding

Auto Mode The mode of operation on a CNC machine used to execute or run the active program

Basic A numerical value used to identify the theoretical exact size, location, or orientation of a part feature. A *basic* dimension is printed inside a rectangular box

Blind Hole A hole that has a depth and does not continue through the workpiece

Blotter A specially designed paper disk attached to the center of a grinding wheel to cushion the flange pressure. Information about the grinding wheel is printed on the blotter

Bond A material that attaches or glues the abrasive particles or grains together in a grinding wheel

Brinell Hardness A measurement scale commonly used when checking the hardness of metals

Canned Cycle A predetermined sequence for performing common CNC machining operations such as hole drilling or rough machining. The command is programmed by a single G code with variables that are entered for the specific operation

Center Punch A hardened metal tool with a conical tip ground to a 90-degree angle. It is used to make or enlarge an indentation on a part's surface prior to drilling

Chamfer The angled edge or corner of a machined part feature, which will provide smooth, burr-free edges and facilitate better assembly of mating components

Climb Milling A method of cutting on a milling machine where the workpiece is fed in the same direction as the rotation of the cutting tool

Collet Chuck A cylindrical-shaped workholding device commonly used on a lathe and is mounted to the spindle nose. It contains a precision ground bore that will accept a matching collet

Conventional Milling A method of cutting on a milling machine where the workpiece is fed in the opposite direction to the rotation of the cutting tool

Counterbore An enlarged, flat-bottomed, cylindrical opening at the top of a hole to allow for the head of a fastener, like a socket-head cap screw to sit flush or below the surface of the part

Countersink a conical or tapered opening, concentric to a hole, to allow for the head of a flathead screw to sit flush or below the surface of the part

Datum A point, plane, or axis from which dimensions are referenced. It is specified on a print by a bold capital letter inside of a square box

Dial Caliper A versatile measuring tool with a graduated scale and a dial that is capable of measuring external, internal, and depth measurements

Dial Indicator A precision measuring tool containing a small movable stylus or contact that measures small distances. The measurement of the distance is indicated on an attached dial face marked with graduations

Drawbar A threaded rod that is inserted through the head and spindle on a vertical milling machine. The threaded end is connected to toolholding devices to secure them

Dresser A device that is traversed across the face of a grinding wheel to dress and true the wheel

Edit Mode The mode of operation on a CNC machine used to make changes to an existing program or type a new program into the MCU's memory

Ferrous Metal A metal containing iron

Flathead Screw A threaded fastener with a cone-shaped head that contains a slot, hex, square, or star-shaped impression for driving the screw

Geometric Dimensioning and Tolerancing A standardized system that utilizes various symbols to communicate part specifications on industrial drawings

Geometry Offset A numerical value stored in the MCU's offset registry that is specific to a particular tool used in the machining process. The value of the offset tells the MCU dimensional data about the specific tool

Independent Chuck A workholding device commonly used on the lathe that has jaws that move independent of each other

Jacobs Chuck A name commonly used when referring to a drill chuck

Jog Mode The mode of operation on a CNC machine that allows the operator to manually move the machine's axes by using the axis buttons or a rotary handwheel

Knee Mill A type of vertical milling machine supported by a vertical column with three axes of table movement

Knurl A straight or diamond-shaped pattern formed on the circumference of a workpiece

Lathe A machine tool that operates by rotating the workpiece while moving a cutting tool into the material to remove specific amounts and produce a desired shape and size

Magnetic Chuck A workholding device commonly used on a surface grinder to secure the workpiece or a secondary workholding device

MDI Manual Data Input. An operation mode on a CNC control that allows the user to enter and execute single commands or a series of single commands

Micrometer A precision measuring tool that operates by rotating a spindle until the desired contact is made between the spindle face, the part being measured, and the anvil face. The measurement is then read from a system of graduated scales

Milling Machine A machine tool that operates by feeding a secured workpiece into a rotating cutting tool

MSDS Material Safety Data Sheet. A standardized document used to communicate information about the hazards associated with chemical, product, or material used in the workplace

Nonferrous A metal containing no iron

OSHA Occupational Safety and Health Administration. The primary federal agency charged with enforcing workplace safety and health regulations

Pedestal Grinder A floor-mounted grinder that uses a grinding wheel is secured to a shaft, which is rotated by a motor

Pinning Occurs when the material removed during filing becomes lodged in the grooves of the file

Prick Punch A hardened metal tool with a conical tip ground to a 60-degree angle. It is used when making layouts to make an indentation on a part's surface to mark intersections of lines and the center of circles or arcs

Range The difference between the highest and lowest values in a set of data

Retention Knob A threaded stud that is secured to the tapered end of a CNC machining center's tool holder. It will secure the tool holder in the machine's spindle when it is gripped by an internal mechanism

Sampling A subgroup selected from a total population from which estimates and predictions can be made

Sampling Plan An outline that is part of a quality control plan and is used to specify what measurements must be taken, on how many parts, and how often

Sine Tool A tool used in workholding, measurement, and layout that is set to a specific angle using gage blocks and a precise reference surface

Spotface A flat circular surface machined on a rough or angled part surface to provide a flat bearing surface for a fastener or washer

Squaring The process of machining the sides of a workpiece perpendicular and parallel to each other and within the part print specifications

Surface Grinder A machine tool that utilizes either a horizontal or vertical spindle, on to which interchangeable grinding wheels are mounted. The workpiece is traversed back and forth against the rotating wheel, and material is removed

Tap A fluted cutting tool that is rotated into a hole to produce threads

Tap Drill The twist drill that corresponds to a tap's specific size and threads per inch

Telescope Gage A measuring tool that has two spring-loaded arms designed to contact the inside diameter of a hole when a locking screw is released. Upon contacting the inside diameter, the locking screw is tightened and the measurement can be obtained by measuring over the contact points with a micrometer

Tramming The process of moving the head of a vertical milling machine until it is precisely aligned to the surface of the table

Tungsten Carbide A material created by combining tungsten and carbide under high heat/pressure. The resulting compound is very hard and wear resistant, making it an ideal material for cutting tools

Twist Drill A cylindrically shaped, fluted cutting tool with cutting edges ground on the tip

Unified National Thread Form A system of standards for ensuring the interchangeability of inch-based threaded fasteners. Size, tolerances, form, and fit are some of the dimensions standardized by this system

Universal Chuck A workholding device commonly used on the lathe that has jaws that move simultaneously to clamp and release from the workpiece

Wear Offset A numerical value entered into the MCU's offset registry that is specific to a particular tool used in the machining process. This offset provides small adjustments to maintain required part dimensions if tool wear becomes a factor during the machining process

Workpiece Coordinates A Cartesian coordinate system that establishes an origin on the workpiece for the purpose of positioning cutting tools and producing a finished part to the correct dimensions

Index

1070 Plain Carbon Steel, 13, 25
2520 Alloy steel, 33, 45

A
Abrasive cutoff saw, 42
Abrasive grain size, 131, 141
Acceptable method, 13, 25
Adjustable parallel, 10, 20
Aluminum oxide dressing stick, 146
Aluminum oxide grinding wheels, 131, 141
American National Standards Institute (ANSI), 1
ANSI. *See* American National Standards Institute

B
Bellmouth, 78
Blind hole, 38, 97
Blocking, 145
Blotters, 133, 144
Bottoming tap, 38–39, 97

C
Candidate registration, 2–3
Canned cycle, 165, 185
Canned tapping cycle, 165, 186
CAR. *See* Credentialing achievement record
Carbide cutting tools, 31, 42
Cast iron, 59, 73, 85, 100
Cast iron shaft, 112, 124
Centerdrill, 39, 66
Centerdrill breakage, 61, 76
Center gage, 120
Center punch, 33, 39, 46, 61, 76
Certification process
    candidate registration, 2–3
    online theory exam process, 4–5
    performance exam process, 3–4
    student/trainee as candidate, 2–3
Chucking reamer, 29
Cleanup process, 7, 11, 17, 23
Climb milling, 41
CNC milling programming, setup, and operations
    actual/desired hole locations, 154, 171

canned cycle, 165, 185
canned tapping cycle, 165, 186
CNC control, position page, 156, 173
countersink tool, 153, 169
Cutter Compensation Interference, 157, 174
cycle start, 170
datum, 156, 172
edge finder, 181
edit mode function, 158, 175
ending/starting the program, 153, 169
flatness symbol, 159, 178
G40 command, 189
G41 command, 167, 188
G codes, 156, 174
Geometric Dimensioning and Tolerancing, 158–159, 176–177
homing, 169
incremental programming code, 157, 175
Jog mode function, 156, 173
linear interpolation, 156, 172
M01 command, 153, 169
M03 code, 165, 185
M08 code, 165, 185
machine control unit, 153, 169
material safety data sheet, 154, 170
offset page, 158, 175
optional stop, 169
preparatory commands, 174
rapid positioning code, 159, 177
retention knob, 175
RPM formula, 159, 178
safety guidelines, 156, 173
spindle code, 160, 178
threaded stud, 157, 175
tolerance, 178
tool length compensation, 163, 182
vertical machining center, 163, 182
workpiece coordinate system, 153, 169
X and *Y* coordinates, 167, 188
CNC turning center, 11, 22
CNC turning programming, setup, and operations
    allowable deviation, 192, 206

constant surface speed, 201, 215, 218
datum, 196, 212
display screen page, 197, 213
Edit mode function, 194, 208
end-return to start code, 194, 209
feed rates, 194, 209
finish turning cycle, 202, 220
G28 code, 202, 220
G40 code, 219
G41 command, 198, 214
G42 code, 218
G70 code, 199, 216, 220
G71 code, 216, 218
G96 code, 193, 207, 218
G99 code, 201, 217
Geometric Dimensioning and Tolerancing, 191, 197, 204, 209, 213
geometry offsets, 191, 205
Jog mode function, 192, 207
linear interpolation, 190, 204
M01 code, 219
M01 command, 195, 210
M03 code, 218
M30 code, 220
modal commands, 216
offset page, 195, 210
pitch diameter, 199, 215
positive rake angle, 198, 215
profile of surface, 194, 209
program stop code, 196, 211
rapid positioning code, 194, 208
right-hand turning tool, 191, 205
rough turning cycle, 199, 201, 216
safety rule, 190, 204
smoothest surface finish, 193, 207
start-up procedure, 190, 204
tolerance, 206
X and Z coordinates, 201–202, 219
Collet, 122
Comparison gage, 8, 18, 90, 105
Compound infeed, 109, 121
Cone-shaped protrusion, 111, 124
Constant surface speed, 201, 215, 218
Control charts, 112, 124
Conventional milling, 95
Coolants, 134, 146, 152